Arctic Encounters

Series Editor
Roger Norum, Cultural Anthropology, University of Oulu, Oulu, Finland

This series brings together cutting-edge scholarship across the social sciences and humanities focusing on this vast and critically important region. Books in the series will present high-calibre, critical insights in an approachable form as a means of unpacking and drawing attention to the multiple meanings and messages embedded in contemporary and historical Arctic social, political, and environmental changes.

Vesa-Pekka Herva · Aki Hakonen ·
Roger Norum · Oula Seitsonen ·
Markus Fjellström

Weirding Landscapes

Arctic Glacier Extinction and Monsters of the
Anthropocene

Vesa-Pekka Herva
Faculty of Humanities
University of Oulu
Oulu, Finland

Aki Hakonen
Faculty of Humanities
University of Oulu
Oulu, Finland

Roger Norum
Cultural Anthropology
University of Oulu
Oulu, Finland

Oula Seitsonen
Faculty of Humanities
University of Oulu
Oulu, Finland

Markus Fjellström
Lund University
Lund, Sweden

ISSN 2730-6488 ISSN 2730-6496 (electronic)
Arctic Encounters
ISBN 978-3-031-85015-8 ISBN 978-3-031-85016-5 (eBook)
https://doi.org/10.1007/978-3-031-85016-5

This work was supported by University of Oulu and Lund University.

© The Editor(s) (if applicable) and The Author(s) 2025, corrected publication 2025. This book is an open access publication.

Open Access This book is licensed under the terms of the Creative Commons Attribution 4.0 International License (http://creativecommons.org/licenses/by/4.0/), which permits use, sharing, adaptation, distribution and reproduction in any medium or format, as long as you give appropriate credit to the original author(s) and the source, provide a link to the Creative Commons license and indicate if changes were made.
The images or other third party material in this book are included in the book's Creative Commons license, unless indicated otherwise in a credit line to the material. If material is not included in the book's Creative Commons license and your intended use is not permitted by statutory regulation or exceeds the permitted use, you will need to obtain permission directly from the copyright holder.
The use of general descriptive names, registered names, trademarks, service marks, etc. in this publication does not imply, even in the absence of a specific statement, that such names are exempt from the relevant protective laws and regulations and therefore free for general use.
The publisher, the authors and the editors are safe to assume that the advice and information in this book are believed to be true and accurate at the date of publication. Neither the publisher nor the authors or the editors give a warranty, expressed or implied, with respect to the material contained herein or for any errors or omissions that may have been made. The publisher remains neutral with regard to jurisdictional claims in published maps and institutional affiliations.

Cover credit: Markus Fjellström

This Palgrave Macmillan imprint is published by the registered company Springer Nature Switzerland AG
The registered company address is: Gewerbestrasse 11, 6330 Cham, Switzerland

If disposing of this product, please recycle the paper.

Dedicated to the memory of Timo-Veli, a very special kind of monster

Series Editor's Foreword

The Arctic. The importance of this region to today's world, and to our planet itself, cannot be overstated. Once viewed as a remote periphery, it is now central to global concerns—environmentally, geopolitically, and scientifically. More than just an emblem and poster child for climate change and human impact, the Arctic serves as a geophysical bellwether and a living laboratory for planetary transformations. It is integral to contemporary debates on energy transitions, resource use (both material and immaterial), and the evolving role of Indigenous peoples in shaping local and global political consciousness. It plays a key role in debates about the pasts, presents, and futures of soil, land, and space. And it has long been embedded in transnational economic and geopolitical considerations, intersecting with European, American, and Asian interests while challenging traditional distinctions, and reflecting prescient anxieties, between centers and peripheries. The recent surge of attention to this part of the world has fueled an unprecedented demand for Arctic research, alongside an expansion of publications, academic programs, and institutions dedicated to the region.

Arctic Encounters is a book series committed to fostering and disseminating cutting-edge scholarship across the sciences and humanities, focusing on this vast and critically important region at the crossroads of Europe, North America, and Asia. The series examines the Arctic's environmental, social, and cultural dimensions—both contemporary and historical—with critical depth and nuance. It embraces a transdisciplinary approach, not only encouraging dialogue across academic disciplines but also engaging perspectives beyond the academy. A core objective of transdisciplinarity is to understand the world in all of its complexity, instead of focusing on any one part of it, or from any singular point of view. By approaching the Arctic as a complex, interconnected space rather than a collection of isolated issues, the series stimulates fresh perspectives in theory, method, and framework, generating novel, out-of-the-box and impactful ways of advancing knowledge.

The series includes both single- and multi-authored monographs as well as edited volumes, while also welcoming less strictly curated forms of expression that are experimental and multi-modal. These may include artistic interventions, collaborative works emerging from research partnerships, and projects co-created with community members outside traditional academic frameworks. By drawing from multiple disciplines and writing traditions, Arctic Encounters seeks to reach not only scholars—including those outside Arctic-specific fields—but also a broader readership of globally engaged citizens concerned with the Arctic's significance in planetary affairs.

Arctic Encounters the series has its origins in Arctic Encounters the project, a HERA-funded research initiative led by Prof. Graham Huggan at the University of Leeds. This collaborative, transnational project sought to bring together distinct perspectives and institutions in the Nordic countries and beyond to address the increasingly important role of the practices and imaginations of tourism and travel writing in fashioning understandings of the European Arctic. Building on this foundation, the book series aims to further "unscramble" Arctic knowledge from the colonial and regionalist frameworks that have historically mapped, constructed and constrained it. With a deliberately wide thematic scope, the series encourages inventive and boundary-pushing scholarship that speaks across disciplines and perspectives, fostering research with environmental, social, and global relevance.

Welcome to the Arctic, and happy reading!

Roger Norum
Department of Cultural
Anthropology
Biodiverse Anthropocenes
Research Programme
Faculty of Humanities
University of Oulu
Oulu, Finland

The original version of the book has been revised: Series editor affiliation has been updated and Series foreword has been included. A correction to this book can be found at https://doi.org/10.1007/978-3-031-85016-5_9

PREFACE

An obscure pulp writer during his lifetime, Howard Phillips Lovecraft (1890–1937) was a founding figure of what came to be known as "weird fiction". Lovecraft's influence is omnipresent across popular culture today, even if his name and direct impact may not be familiar to casual consumers. The American author Matt Ruff's bestselling novel *Lovecraft Country* (2016), and the recent television adaptation based on it (2020), presented Lovecraftian themes and monsters to wide audiences, juxtaposing them with the racist terrors of America in the 1950s. But Lovecraft and his legacies had been widely engaged with for decades even if this took place primarily in subculture circles. Thus, for instance, Metallica released an instrumental piece called "Call of Ktulu" on their 1984 trash metal album, *Ride the Lightning*. The piece referenced Lovecraft's most famous story and monster, "The Call of Cthulhu" (1928); Metallica's later 1986 album, *Master of Puppets*, featured a song called "The Thing That Should Not Be". This song title effectively captures what the "weird" is about. Over the 2000s, the weird has come to feature in various scholarly and cultural discourses, most prominently as relates to the world of the Anthropocene, ever weird in diverse forms and senses.

This book attempts to mediate between the often hyperbolically theoretical writing on the Anthropocene and diverse, quotidian senses that there are all kinds of "things that should not be"—or things that feel "wrong"—in our contemporary world and in our experiences of that world. Indeed, this very book is a thing that should not be in the literal sense: we hardly anticipated an entire book to emerge from a week's bout of fieldwork on the lingering remains of a perennial ice and snow field in Europe's High North. It didn't even cross our minds to write a book when we boarded a helicopter in the diminutive village of Kilpisjärvi in Finnish Lapland/Sápmi on our way to our research area in August 2022, nor did we think of doing it when we returned to Kilpisjärvi a few days later. Something did take place upon our return, however. Independently, each of the five of us felt that the field trip had triggered something in us: something weird and monstrous had been released during our few days in that northern "wilderness". It was perhaps not quite like the release of Cthulhu from its submerged city of R'lyeh, as Lovecraft memorably described, but the trip released *something* weird in our minds and in the ways we perceived that world of the field, and the world we had returned to. As we considered this further over the coming months, we came upon the notion of "weirding"—that is, using weird fiction for the purposes of seeing the world and thinking about it from a particular angle.

This book, of course, did not write itself. But the writing process—and its end result—was something markedly different from the kind of academic writing we had prior engaged in. It is perhaps telling that when we began to write something about our engagements with those weird high northern lands, one of the first things we did was retrofit an academic (and Arctic) version of the marketing poster for Quentin Tarantino's film *The Hateful Eight* (2015). This might have remained but a hazy idea, supported by many pages of notes in Vesa-Pekka's notebook, had it not been for the happy coincidence that (months after our return) María Ximena Senatore and David González-Álvarez approached Vesa-Pekka with an invitation to contribute to a special issue on the archaeology in remote lands. We thought that this was a promising platform for developing a reflective piece on the fieldwork, and indeed the guest editors encouraged us to do so.

Only, things got weird again. First, it turned out impossible for us to *contain* the piece that we had agreed to write. This is admittedly hardly a unique experience for academic authors; overwriting and submitting bloated text (or rambling in a conference presentation well over our allotted 15 minutes) is nearly a requirement of our métier. But this was different. When we wrote, it felt, every so often, as though something had been released, and that something just kept pouring out. The problem, in concrete terms, was that the text soon surpassed the word count limits of any ordinary journal article. Eventually, however, we mustered a submittable manuscript, one that was in fact favourably received by both the peer reviewers and guest editors. But the planned special issue ultimately did not come together. After this discouraging news, we found that we were unwilling to discard the text that by then had begun to feel as though it was itself a living and breathing entity. Substantially cutting it back at that point would have killed our attempt to take a "weirding perspective" on the Anthropocene Arctic. Although the manuscript was *very* long at this point, we still felt that much was still missing and decided to pursue it as book. After we made that decision, it took us about one year—from the autumn 2023 to autumn of 2024—to turn our original text into a book manuscript and as we are about to press submit, we are delighted to have taken that road.

Oulu, Finland Vesa-Pekka Herva
Tampere, Finland Aki Hakonen
Oulu, Finland Roger Norum
Helsinki, Finland Oula Seitsonen
Lund, Sweden Markus Fjellström

Acknowledgements

Given that this publication has gone through multiple instantiations, we have drawn on a wide range of core resources to make this book happen. Foremost, we want to acknowledge and thank the Arctic Avenue and the Swedish-Finnish Cultural Foundation for financing our initial field project "Archaeological survey of snow patches in the Háldi-Ritničohkka region, Sápmi" (PIs Oula Seitsonen and Markus Fjellström). We are extremely grateful to the Finnish Border Guard, National Board of Forestry Metsähallitus, the Sámi Museum Siida, and the reindeer cooperation chief Juha Tornensis for their continuing help and assistance with various practicalities of our being in the North. Sanna, Sohvi, Elsa and Elvi Seitsonen participated in the fieldwork on Háldi in 2023 that partly contributed to the material for this book. Telia Company kindly let us stay in and make use of their old maintenance cabins on Ritničohkka during our fieldwork in 2022. We thank the several anonymous reviewers, as well as Tuuli Matila, María Ximena Senatore and David González-Álvarez, for encouragement, reading, commenting, and making useful and insightful suggestions on previous drafts of the manuscript at different stages. Sophie Schlondorff performed a marvellous task in wonderfully proofreading the final manuscript, while Jenna Autio, Saija Kostamo and Joel Loukkola, top students on Andrew Pattison's language editing course at Oulu, gave us extremely insightful comments and edits on a latter draft of the book.

The original work in Ritničohkka and the writing of this book have been carried out as part of several research projects, and was a sister project for the project "Archaeological Surveys of Melting Glaciers and Snow Patches in Swedish Sápmi" (PIs Kerstin Lidén and Markus Fjellström, Stiftelsen Marcus och Amalia Wallenbergs minnesfond grant number MAW 2020.0122). Much of the analytical work reported in this book took place in the context of the Research Council of Finland-funded project "Extractive Industries as Engagement with the Extraordinary Subterranean: Culture, Heritage and Impact of Resource Extraction in Northernmost Europe" (PI Vesa-Pekka Herva, funding decision number 339483). The fieldwork and writing was also assisted by other contributions from other major projects, including "Pohjoisen päätepisteet: sosioekologisesti kestävän kehittämisen ja liikkumisen nykyisyys ja tulevaisuudet arktisessa Fennoskandiassa" (PI Roger Norum, Research Council of Finland, funding decision number 339423) and by the ERA.Net RUS Plus programme under award number RUS_ST2019-055 for the consortium "CONTOURS: Conservation, Tourism, Remoteness" (PI Roger Norum, Research Council of Finland, funding decision number 344717). It has also been latterly inspired and supported by CHANSE and HERA (RURALEX), through the Research Council of Finland, via the development of the project "Knowledge in Crisis: The Dynamics of Environmental Expertise amidst Rural Change". Additionally, the research and the writing benefited from support from "Eerie Arctic Aviation: Impacts and Cultural Legacies of 20th Century Military Aviation in the Circumpolar Zone" (PI Oula Seitsonen, Research Council of Finland, funding decision number 363043) and "Contemporary Archaeological Perspective on the Inequality of the Welfare Society" (PI Oula Seitsonen, Kone Foundation, funding decision number 202203221). Oula's work was also partly supported by the Finnish Committee for Public Information and the Association of Finnish Non-fiction Writers. Aki had no funding.

We are, finally, extremely grateful to Prof Paula Rossi, Dean of the Faculty of Humanities at the University of Oulu, Finland, for her untiring support of our research ideas, however weird they might initially sound. And we owe great thanks to the Lund University Library, Sweden, for funding the Open Access costs for this book, along with contribution towards such costs by the "Biodiverse Anthropocenes" Profit Profiling Area programme, University of Oulu (PIs Marko Mutanen and Roger Norum).

Contents

1 Introduction 1
   1.1 An Impression 1
   1.2 Real and Imagined Arctic 4
   1.3 "Fielding" Ritničohkka 8
   References 11

2 Weirding and Fielding the World: Hows and Whys 15
   2.1 Tales of Two Landscapes: Monstrous Relations and Resonances 15
   2.2 Weirding 18
   2.3 Monsters and the Monstrous 24
   2.4 Bodies on the Move 27
   2.5 Making Sense of the (Anthropocene) World Through Fieldwork 28
   2.6 Fielding in Practice 32
   References 36

3 The "Glacier" 45
   3.1 A Flying Transition 47
   3.2 The Fjells of the North: Political Geographies, Mindscapes and Mythical Resonances 51

3.3	The Discovery of the Glacier	55
3.4	What Is a Glacier?	60
3.5	Weirded Rocks of a Post-glacial World	63
3.6	The Horror of Rock and the Emotions of Deep Time	69
References		74

4 Mountain Beings — 81
- 4.1 Nearing the Monstrous, Weirding Description — 82
- 4.2 A Sea of Stones — 83
- 4.3 Mountain Beings and Mineral Evolution — 86
- 4.4 Absences and Presences of Life, Death and Sentience — 89
- 4.5 Reindeer, Mosquitoes and the World on the Move — 92
- References — 96

5 At the Basecamp — 99
- 5.1 First Encounter — 99
- 5.2 A Mobile Dead Zone: Roger's Account — 104
- 5.3 Spiritual Communications — 107
- 5.4 Uncanny Engagements with Technology — 110
- 5.5 Faecal Action — 115
- 5.6 The Cabin at the End of the (Other) World — 118
- References — 121

6 The Fjell in the Cloud — 127
- 6.1 "This Is Where You Find Cthulhu" — 127
- 6.2 All the Shades of Grey — 129
- 6.3 Disorientation and Labyrinths of Stone — 132
- 6.4 Manifest and Hidden Dimensions — 135
- 6.5 Enminded Bodies and Other Instruments for Accessing Invisible Worlds — 138
- 6.6 Air and Smells as Intangible Artefacts — 142
- 6.7 An Escape Through a Deceiving Land — 144
- References — 148

7 Gear Shift: Hiking and Being in the North — 153
- 7.1 Public Wilderness Huts in Finland — 154
- 7.2 Hallucinatory and Spectral Huts — 156
- 7.3 Human Encounters — 158

7.4	*Survival Tech*	160
7.5	*Hauling a Mobile Home*	162
7.6	*Cultures of Walking*	165
7.7	*Artefacts of the Foot*	168
7.8	*"Huiputus" and Other Tales of Conquest*	172
7.9	*"Big-Man" Legends of the Trail*	174
7.10	*The Sociality of Solitude*	178
	References	183

8 Conclusion: Monstrous Worlds — 187
References — 192

Correction to: Weirding Landscapes — C1

Correction to: The "Glacier" — C3

Index — 193

About the Authors

Vesa-Pekka Herva is Professor of Archaeology at the University of Oulu, Finland. His research interests encompass material culture, human-environment relations and cosmology in the European High North from the Neolithic to the present day. He has led several major research projects, including "Extractive Industries as Engagement with the Extraordinary Subterranean: Culture, Heritage and Impact of Resource Extraction in Northernmost Europe" (Research Council of Finland, 2021–2025). Herva has published on a wide range of topics from the dynamics of Neolithization to the difficult heritage of the Second World War and the uses of cultural heritage in Lapland Christmas tourism. He is the author, with Antti Lahelma, of *Northern Archaeologies and Cosmologies: A Relational View* (Routledge, 2020).

Aki Hakonen is an archaeologist, Ph.D., and a fledgling science writer. He is currently a post-doctoral researcher at the University of Oulu, Finland, in the project "Eerie Arctic Aviation: Impacts and Cultural Legacies of 20th Century Military Aviation in the Circumpolar Zone" (Research Council of Finland, 2024–2028). His prior research has focused on material and immaterial networks of prehistoric Fennoscandia.

Roger Norum is Associate Professor of Cultural Anthropology at the University of Oulu, Finland. He received his doctorate from the University of Oxford in 2015 with a thesis on the sociality and temporality of skilled labour migrants in Kathmandu, Nepal. His current research across the Arctic and South/South-east Asia focuses on topics of the place of mobility, sound, and sociality in human-environment relations. He is PI of the research projects "RURALEX: Knowledge in Crisis—The Dynamics of Environmental Expertise amidst Rural Change" (CHANSE-HERA, 2025–28) and "FORbEST: Safeguarding Carbon and Biodiversity across European Forest Ecosystems thru Multi-actor Innovation" (Horizon Europe, 2025–29). He is the co-author, with Alejandro Reig, of *Migrantes* (Ekaré, 2019).

Oula Seitsonen, Sakarin-Pentin Ilarin Oula, is Associate Professor (Title of Docent) at the University of Oulu, Finland. He is both a geographer and an archaeologist by training and his research covers diverse themes ranging from the Stone Age of East Africa and Mongolia, to archaeology and history of reindeer domestication and herding, and to recent past Arctic conflict heritage. Seitsonen has acted as the Chair in Finnish Studies at the Lakehead University, Canada (2022–2023), and is currently a Clare Hall Fellow at the University of Cambridge marvelling at the links between the Second World War German activities in Finland and the international Holocaust discourse (2024–2025). He is the PI of the projects "Contemporary Archaeological Perspective on the Inequality of the Welfare Society" (Kone Foundation, 2022–2027) and "Eerie Arctic Aviation: Impacts and Cultural Legacies of 20th Century Military Aviation in the Circumpolar Zone" (Research Council of Finland, 2024–2028). Seitsonen is also the author of the first monograph dealing with the Second World War materialities in Finland from a theoretically informed perspective, *Archaeologies of Hitler's Arctic War: Heritage of the Second World War German Military Presence in Finnish Lapland* (Routledge, 2021).

Markus Fjellström is Post-doctoral Researcher in Archaeology at Lund University where he studies Late Paleolithic and Early Mesolithic reindeer mobility in southern Scandinavia using stable isotope and aDNA analysis. He has a Ph.D. in Archaeological Science from the Archaeological Research Laboratory at Stockholm University, to which he also is affiliated, focused on ancient human and animal diets, mobility, and environmental changes through the use of stable isotope analysis. Recently, at the University of Oulu, his research has explored reindeer domestication in northern Fennoscandia, Sápmi, and the Arctic by combining stable isotope data with archaeological and historical sources. Since 2015 he has also been involved in developing glacial archaeological approaches and surveying melting glaciers and snow patches in Swedish and Finnish Sápmi. Additionally, he collaborates with the Silvermuseet/INSARC in Arjeplog, examining reindeer mobility and related Sámi taxland research questions.

List of Figures

Fig. 1.1	The helicopter taking off from the summit of Ritničohkka on Day 1	2
Fig. 1.2	The perfect depiction of a magical, enchanted and enchanting Arctic Lapland: Aurora borealis dancing above the sacred Sána Mountain at Kilpisjärvi (Wikimedia Commons/WikiLucas00/CC BY-SA 4.0)	5
Fig. 1.3	An Arctic Ocean cruise ship leaving the Port of Longyearbyen in Svalbard; in the foreground, reminders of the remote archipelago's over century-long commercial mining heritage (Wikimedia Commons/Zairon/CC BY-SA 4.0)	8
Fig. 1.4	Location of Ritničohkka and other relevant locations in northwesternmost Lapland: (1) Ritničohkka basecamp; (2) Háldi summit; (3) Kilpisjärvi; (4) Hiking trail; (5) Háldijávri cabin (Background maps from Google Earth and National Land Survey of Finland, orthophoto dated to summer 2012 shows the general summertime extent of the Ritničohkka snowfield as it was until the early 2000s)	9
Fig. 2.1	View of the abandoned mining landscape at Hannukainen with the Ylläs ski resort in the background (left), and barren tundra at Ritničohkka (right)	16
Fig. 2.2	View of the infinite Arctic tundra from the summit of Ritničohkka	31

Fig. 3.1　A bird's eye view of our expedition's journey from the south-east. The extent of the snowfield in the aerial imagery is as it was in 2012 (National Land Survey of Finland, orthophoto dated 2012 and 2 m resolution digital terrain model [DTM] and Norwegian Mapping Authority, 1 m DTM). The light grey shading shows our route as recorded by GPS navigators　46

Fig. 3.2　Soaring over the barren reindeer country　50

Fig. 3.3　View of the Háldi summit, the highest point in Finland　51

Fig. 3.4　With each uncertain step, the team scavenged the remains of the vaunted glacier　57

Fig. 3.5　Another glacial refuge not long for this world. Note the difference between the recently uncovered vegetation-free light grey rocks and the darker shades in the distance　58

Fig. 3.6　Markus exploring the last subglacial　58

Fig. 3.7　A glacier that once was: remains of Mer de Glace (Eng. "Sea of Ice") in Chamonix, French Alps. Notice the sign showing the 1990 surface level of the glacier. Even in the early 2000s, the valley was still filled with glacial ice to the upper limits shown by the open soil, but since then it has been shrinking at an ever-increasing speed of some 40 metres per year, losing about 80 metres in thickness. At the bottom of the steep ravine, one can see the pitiful remains of the once massive ice core　62

Fig. 3.8　The pockmarked face of a melting snow patch, jassa, on the slope of Gieddečohkka (Fi. Etu-Halti), with the summit of Háldi on the far right　63

Fig. 3.9　Weird banded gneiss of the nappe from 420+ million years ago, reaching for the sky　66

Fig. 3.10　A close-up of an eye pattern disturbingly oozing with quartz. Notice the fluidly twisted stratigraphy frozen in time　67

Fig. 3.11　Geological peculiarities inspired the attuned mind to come up with fantastic explanations, such as ancient and unknown lifeforms　67

Fig. 3.12　Markus getting unsuspectingly crushed by Ymir's ghost　72

Fig. 3.13　Stepping onto the abyss of time. A recently exposed subglacial surface, with twisted mineral sheets characteristic of gneiss, reveals mesmerizing metamorphoses dating from the Silurian Age, flatting the vastness of time into a surface　74

Fig. 4.1	Things (and people) out of place	83
Fig. 4.2	Cliff walls erected by geological forces present imposing surfaces with indescribable, "wrong" geometries. A melting ice patch lies mournfully at the feet of the high escarpment that rises abruptly for over 20 unscalable metres	84
Fig. 4.3	Mountain beings. Roger and Vesa-Pekka pondering on a weird landscape	88
Fig. 4.4	Minerals evolving, with many forms of lichen extracting minerals from the rock and renewing the world. Moss and lichen are slowly creeping towards a freshly bleached bone	89
Fig. 4.5	Fresh looking yet centuries-old reindeer bones recently uncovered from under the ice and snow at Ritničohkka	93
Fig. 5.1	The cabins at sunset on the first evening	100
Fig. 5.2	Roger attempts to open the lock while others carry bags and bundles of gear from the landing zone. On the right, one of the cairns built next to the cabin	101
Fig. 5.3	There may have been a reason why the door to the cabin was bolted tightly shut	102
Fig. 5.4	The emergency rations in the cabin turned out to be several decades past their expiration date	103
Fig. 5.5	Clothes hanging inside the cabin. Electric cables and bunk beds tell stories of the cabin's glory days; the thoroughly mouldy mattresses were thrown into the storage shed	104
Fig. 5.6	Blood-red midnight sun hanging low behind a wisp of clouds. Silhouettes of the cairns in the foreground	111
Fig. 5.7	Movie matinée in the cold cabin while waiting for the bad weather to pass	112
Fig. 5.8	Stone cairns in the mist on the fjell's high point, adjacent to the deceitfully inhabitable-looking shed	114
Fig. 6.1	A monstrous landscape and an imprisonment of perception in the mist, at once awe-inspiring and horrific	128
Fig. 6.2	Grey hides and reveals the last of the Ritničohkka snow and ice patches	131
Fig. 6.3	Subjective and objective merge and twist in a messy landscape	146
Fig. 6.4	Skies are alive	147
Fig. 7.1	Lake Háldijávri wilderness hut	154

Fig. 7.2　Our five-person team may have had more luggage with us than your average hiker. Various backpacks, duffle bags, bundles and rucksacks of assorted fieldwork gear, with Roger's green rolling suitcase in the middle, unloaded from the helicopter on the summit of Ritničohkka　163

Fig. 7.3　A Tanzanian Maasai friend of Oula, Israel ole Mollel of Engaruka, walking across the savannah in his car-tire flip-flops while assisting in archaeological fieldwork, with Mount Kilimanjaro rising in the distance　168

Fig. 7.4　An abandoned shoe entangled with its environment (left), featuring its own lichen cover, which, under the microscope, reveals a micro-landscape that parallels the barren valleys and peaks of the surrounding land (right)　171

Fig. 7.5　President Kekkonen (on the right) and his entourage visiting a Sámi tent in Gáijohaš village at Ávžžášjávri on his skiing trip in 1977 (Finnish Heritage Agency HK7966:34/Einar Närhisalo/CC BY 4.0)　175

Fig. 7.6　A view of Čáivárri　177

Fig. 7.7　A view of Háldi (on the far left) and the slope of Ritničohkka (on the far right), with Oula racing to an outhouse which postdates Kekkonen at the Lake Háldijávri hut　177

Fig. 7.8　Roger engaged in "automatic writing" during a writing retreat for this book in Helsinki　183

Fig. 8.1　Farewell to Ritničohkka, its two summits rising above the clouds, as seen from Háldi　188

CHAPTER 1

Introduction

1.1 An Impression

The dulled patina on the olive-black shell of the noise-cancelling headsets suggested they had been part of the long-serving helicopter's gear for some time. Inside the cockpit, a mobile phone panned the horizon. Radiating off the screen, the image resembled the opening scene of a 1970s blockbuster. Outside, an otherworldly boulder-ridden landscape of valleys and streams whisked by, the shadow of the chopper skimming along the surface. The roar of the turbine engine must have echoed in every nook and crevice, but the headsets guarded us from the noise, mellowing it down to a low electric purr. Levitating in the clear blue sky above Finnish Lapland's furthest reaches, we found ourselves at once acutely present in the real world, yet vaguely, uncannily detached from reality (Fig. 1.1).

In the observer's seat next to the pilot, Oula was fiddling with the all-important key in his pocket, scanning the surroundings with a mischievous look of excitement. Behind him, Markus updated a mental checklist of the team's scientific objectives, while Vesa-Pekka stared out towards the horizon, pondering the dimensional warp of our flight up from Kilpisjärvi. Roger was augmenting reality by testing the different image filters of his iPhone, with Aki dreamily projecting himself among the rocky oases and rushing streams below.

A lonely telecommunications antenna coming into view in the distance announced our arrival.

Fig. 1.1 The helicopter taking off from the summit of Ritničohkka on Day 1

On top of the mountain, the view extended to the far horizon. Sitting on desolate rocks, amidst our excessive gear, we observed forsakenly as the helicopter ascended, leaving us to our own devices. Disappearing into the distance, back down the valley from which it came, the echo of the chopping beat faded. One by one, we rose up and turned to face the rundown cabin and inactive antenna. According to our plan, this would serve as our base camp for the next five days.

On the fifth day, outside a completely different wilderness cabin, a crowd of ten people waited in the wind and drizzle in anticipation. As the clouds above were getting darker again, the pickup was far from guaranteed. Oula—his forlorn key and a sealed bag of prehistoric stone flakes in his pocket—checked the time yet again, anxious yet sad to leave the field, his natural habitat. Vesa-Pekka, holding a notebook scribbled to the brim with insights on invisible agencies, was lighting another cigarette, fed up with the waiting and the rain and the chill, as well as the sociable antics of Roger, who, after playfully bantering with hikers with whom we had unexpectedly come to share the hut, drifted off in a futile search for bars on his phone. Aki, boots wet and thinking of home, was hesitantly looking for excuses to stay, dismayed at the impending breakup of this newly formed, small community. Markus shuffled the innards of his backpack around to make sure none of the recovered bone samples were getting crushed, unsure of what to make of it all.

After several false echoes, the familiar faint din of whirring became audible. The helicopter appeared below the low cloud ceiling over the adjacent lake, navigating a narrow topographic tunnel lined by the steep valley slopes. It approached the valley's dead-end, where the wilderness hut and prescribed landing site stood. From here, a lone hiking trail ascended towards the highest point of Finland and into the clouds that shrouded the surrounding fjells.

Goodbyes were said and empty promises to stay in touch were given. With the team onboard, the chopper ascended and skidded across the lake, leaving the wilderness hut and the dwindled crowd behind.

"So, how'd it go?", the pilot asked, and for a while, there was no answer.

* * *

What started out as a fairly normal research paper on the strangeness of scientific fieldwork in the north gained a life of its own. A few months after the journey in August 2022, we began drafting a narrative, which eventually ballooned beyond 18,000 words, of an indefinite kind that one befuddled reader described as "a metaphysical adventure story". In it, we sought ways to approach the otherworldliness we had experienced, seeking to define it in ways that would help us answer the lingering question: *what was that?*

It took some time to recognize that this was in fact not a paper but a book, which we wrote mainly in autumn 2023 and spring 2024. In hindsight, the narrative format, at odds with accustomed academic writing, had an unexpected impact on the process, permitting the free flow of associations and experimentations. In putting together our manuscript, we began to recognize a vague resemblance to Edgar Allan Poe's renowned *The Narrative of Arthur Gordon Pym of Nantucket* (1838). Poe's *Narrative* starts out as a seemingly light-hearted adventure story, but turns from ominous to macabre—including accounts of cannibalism—to fantastic, until it finally unravels at the end. We don't pretend to be as original as Poe, and our narrative is not fiction, but we do draw inspiration from the tradition of fiction—horror and, in particular, what is known today as "weird fiction"—whose foundations Poe laid and that inspired later generations of writers. The works of weird fiction by Howard Phillips Lovecraft especially came to play a significant role in processing our fieldwork on Ritničohkka and its broader implications for making sense of the monstrous Anthropocene worlds.

1.2 Real and Imagined Arctic

"Lapland, the home of lap dance", muses one character in the British sitcom *Men Behaving Badly* from the early 1990s. While this joke is a passing wordplay in the series, it resonates with a long tradition of imagining Lapland, in northernmost Fennoscandia, as a mentally and physically peripheral fantasyland. Indeed, coincidentally or not, it was only some years prior, in the mid-1980s, that Finnish Lapland had emerged as an exclusive Christmas tourism destination that was successfully marketed as the home of Santa Claus. Wealthy Brits were flown in for Christmastime day-trips to Lapland—specifically to the town of Rovaniemi on the Arctic Circle—aboard the extravagant supersonic Concorde jet. This was a very special vehicle for taking tourists to an exotic (sub-)Arctic world.

Capitalizing on centuries-old ideas and perceptions of the enchanted and magical High North, the imagery informing Lapland Christmas tourism has been circulating in Europe, in particular since the late seventeenth century, following the Western European "discovery" of this unfamiliar and strange remote land. These early modern perceptions were, in turn, founded on ancient and classical ideas about the far North (Herva et al. 2020). The etymology of Lapland is old and muddled, meaning "Land of the Lapps". The name "Lapp" seems to have been applied originally from the outside to denote the otherness of local people leading a lifestyle primarily of foraging still prevalent in the north until a few centuries ago (Aikio 2022; Lehtola 2009). In its modern use, Lapp is a derogatory name for the Sámi who inhabit the far north of the Fennoscandian peninsula.

The Sámi are considered the last indigenous people of the European Union, loosely adjoined by cultural and linguistic ties. The transnational, cross-border homeland of the Sámi people is called Sápmi. Unrecognized by state governments, it extends over some 400,000 sparsely settled square kilometres, from the Arctic coastline and snow-peaked mountains of Norway and Sweden, via the highland tundra and stunted boreal forests of Northern Finland, and across the scraggly Kola Peninsula of Northwest Russia. In this land, traditional, local Sámi settlements, land use and reindeer pastures intermingle with colonialist mining towns and tourism hotspots.

Different aspects of these real and imagined high northern lands have fascinated southern travellers for many centuries (Fig. 1.2). And, while we ourselves are no strangers to this northern land, it still manages to feel special, magical and evocative to us every time we find ourselves there. This has to do with various factors that range from personal and cultural imaginaries to the actual characteristics of Lapland landscapes, which may initially appear merely as a vast, picturesque and empty (sub-) Arctic wilderness of taiga, tundra and peatland that only gradually reveal more of their richness.

Despite our relative familiarity with Lapland compared to casual, city-break tourists, this land and its landscapes continue to impress, surprise and spellbind when one approaches them with an open and curious mind. We have been aware of the land's affective power for many years—some of us indeed since our childhoods—but only recently have we begun

Fig. 1.2 The perfect depiction of a magical, enchanted and enchanting Arctic Lapland: Aurora borealis dancing above the sacred Sána Mountain at Kilpisjärvi (Wikimedia Commons/WikiLucas00/CC BY-SA 4.0)

to engage with its deeper significance and the implications this holds both for carrying out archaeological and anthropological fieldwork in the region and for understanding the manifold aspects of human-environment relations in the age of the Anthropocene.

Some of these issues surfaced somewhat prominently in August 2022, when we conducted an archaeological survey—the fieldwork of this book—on the fjell (mountain) of Ritničohkka in the very northwest corner of Finnish Lapland, deep in the Sápmi heartland. This comparatively diminutive peak, on an international scale, is situated in the highest region of Finland, known as the "Upper end" (Fi. Yliperä) in the local vernacular. All Finnish peaks exceeding 1000-meter elevation are found in this area across the mountains from the island city of Troms in Northern Norway. Lapland, and the Arctic in general, has attracted scientists, explorers and travellers since early modern times. Today, it has become such a climate and polar tourism hotspot that its sensitive socio-ecological systems are facing important sustainability challenges. The fact that it is also home to the European Union's only indigenous people, the Sámi, presents a cornucopia of reasons for scholars and sojourners alike to seek out this part of the world. Lapland, like other parts of the Arctic, is in the midst of unpredictable transitions focusing on fossil-free energy, environmental conservation and sustainable tourism (Hovelsrud et al. 2020).

For millennia, the far-off Arctic was seen in many imaginaries as a mythical place populated by people living in a fabulous landscape of eternal spring, sunlight and warmth (e.g. Davidson 2005). For those who had never visited the Arctic, it was understood as a land of enlightenment, devoid of shadow and full of possibility—a land of happiness where one could overcome the decadence and malady of the modern individual (Davidson 2005). Today, the region continues to be mythologized in many ways as an unknown, inaccessible and forbidden land, a remote, wild and often "othered" frontier.

While "the Arctic" is a broad topographical description we use to orient ourselves, it is frequently referred to in a way that essentializes this diverse and contested space, reducing the complex geopolitical, (trans)national, cultural and linguistic distinctions that define it. But the alterities often ascribed to this part of the world are both multifarious and socially, culturally and historically embedded. The Arctic's multiple landscapes—and its aesthetic, cultural, political, economic and scientific interests—are all inexorably intermeshed with one another.

During the age of discovery, the Arctic served as a space for the ideological projections of the European (usually male) individual. It has also played the role of geographical, topological muse for thought and experimentation, for creativity and artistry, for collaboration and for adventure—for artists, writers, thinkers, scholars, explorers and travellers. Journeys to the North have, over time, been framed as odysseys towards the self and truth, movements mobilized by the poetic potential of exploration, both to the ends of the Earth and the innards of the soul. The landscapes of the Arctic—reluctant, disorienting, alienating; remote, lonely, vulnerable—can hold the power to bridge boundaries, to contest categories and to devise novel discourses.

It should perhaps come as little surprise then that the first tourists to the Arctic were scientists: men of empires, empowered and emboldened by their monarchs to survey the lands of the North and return home with knowledge, spoils and stories. But at the same time, the Arctic that drew seafaring explorers and adventurers a thousand years ago is not strictly the same Arctic that we visit today. For one, the Ryanair-age of ultra-low-cost airlines makes it inexpensive to get there quickly; at the same time, the melting and vanishing ice makes it feel all the more urgent to do so, while simultaneously fuelling global warming.

Indeed, many hotspots of climate change across the planet have long drawn both tourists interested in seeing how the planet is changing and scientists interested in documenting, or halting, such change. Such seemingly divergent groups can in many ways compete for access to threatened or precarious environmental spaces. The Arctic being a focal point for environmental change (Demiroglu & Hall 2020), with rates of snow and ice loss among the planet's highest (Pörtner et al. 2022), the region increasingly draws more and more people who come to see it before it melts (Huggan and Norum 2015), turning it into one of the planet's "last chance" destinations (Müller 2022; Varnajot and Saarinen 2021) (Fig. 1.3).

Most often, tourists are consumers of place and of experience, whereas scientists are seen as producers of knowledge (Glasberg 2012). This book takes as its point of departure a short archaeological survey carried out by five researchers–scientists, to be sure, but also tourists. By following the movements of their tour, it considers their relationships and the meanings and contexts they bring to the landscapes that they are affected by as data-obsessed researchers, yes, but also as curious adventurers of the North. Considering such roles embodied within the same individuals can give a

Fig. 1.3 An Arctic Ocean cruise ship leaving the Port of Longyearbyen in Svalbard; in the foreground, reminders of the remote archipelago's over century-long commercial mining heritage (Wikimedia Commons/Zairon/CC BY-SA 4.0)

better understanding of the entangled realities of how imagination and knowledge, and labour and leisure, are intertwined (Zhang et al. forthcoming). Indeed, their connections bring to light compelling questions about sustainability, environmental impact and consumption of planetary resources, in part because science and tourism do compete over access to important environmental spaces (Rutty et al. 2015).

1.3 "Fielding" Ritničohkka

The name Ritničohkka is from the North Sámi language and means a "crown snow-load peak". The name references the heavy snowfall typical to this fjell and the accumulation of thick snow cover that perpetually bedecks it—and partly explains why a persistent field of snow and ice, marked as a "perennial glacier" on multiple Finnish government topographical survey maps, has formed in this particular location. Ritničohkka snowfield, a solid field of snow and ice, is a unique glacier-like feature in Finland, which provided the impetus for our survey (Fig. 1.4). With glaciers melting all around the world, including along the Scandinavian mountain range, pertinent archaeological material formerly encased in ice is in danger of decomposition. Ritničohkka did not disappoint us, but our time on—or with—it turned out to be different from what we initially planned and expected.

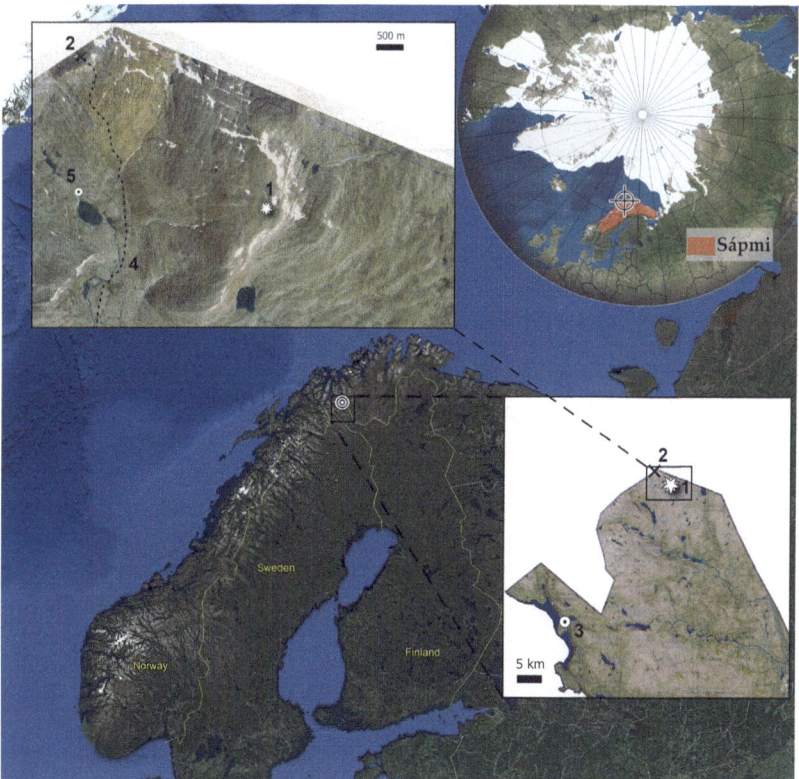

Fig. 1.4 Location of Ritničohkka and other relevant locations in northwesternmost Lapland: (1) Ritničohkka basecamp; (2) Háldi summit; (3) Kilpisjärvi; (4) Hiking trail; (5) Háldijávri cabin (Background maps from Google Earth and National Land Survey of Finland, orthophoto dated to summer 2012 shows the general summertime extent of the Ritničohkka snowfield as it was until the early 2000s)

While we made interesting archaeological discoveries during our survey, this book is primarily about broader issues related to the practices and perceptions of doing archaeological fieldwork in a remote land—encountering, describing, feeling and knowing a "strange" land. Fieldwork is often seen to consist of objective documentation and data collecting, but it is also a particular, subjective form of engaging and interacting with land, which in turn involves much more than just "practical" activities.

Fieldwork—or rather "fielding", to highlight its active, open-ended, experiential, experimental and affective nature (Paphitis et al. 2021)—is a form of encountering the world that pays attention to its myriad, layered and entangled natural, socio-cultural and spiritual dimensions. Fielding becomes a medium for reflecting on our place in, and contextualizing our relationships with, the surrounding world and with its diverse other-than-human constituents and co-inhabitants, as well as its situated nature emplaced between pasts and presents, presences and absences.

A few years ago, we developed an interest in archaeological and anthropological fieldwork as a practice, specifically as a particular way of encountering and relating with the surrounding world (Paphitis et al. 2021). For instance, we pondered on what the various animate and inanimate, tangible and intangible constituents of landscapes "do" in the context of fieldwork—how they afford or inhibit movement, perception and thinking. These considerations unfolded in relation to our more general and long-lasting interest in relational ontologies and epistemologies, both theoretically and as a characteristic of northern cultures (Ingold 2000, 2010; Herva and Lahelma 2020). The basic tenet of relational thinking, as we have engaged with it, is quite simple: relationships between entities define what entities are and what properties and qualities they have. Consequently, knowing a relationally constituted (highly dynamic) reality requires situational awareness of relationships between different entities.

A more difficult question is how to employ relational thinking in real-life cases and situations, such as landscape characterization and documentation in the context of fieldwork. At the same time, the contemporary Anthropocene world is populated not only by plants, animals and humans, but also diverse more-than-human—even monstrous—entities and forces that are partly generated by humans but exist and operate in the world in ways that are both beyond human control, powerful, elusive and shapeless, and yet very real and omnipresent, shaping human and non-human worlds in decisive but often invisible ways. These include hyperobject entities such as climate change, pollution, capital, digital technology and algorithms (Morton 2013)—monsters of the Anthropocene.

We ventured to Ritničohkka originally to carry out a conventional—albeit exploratory—archaeological survey. Very soon upon our arrival, however, we realized that Ritničohkka was working on us too, which introduced a new and unexpected angle to our fielding. Ritničohkka had a strong impact on us and brought to the surface our earlier ponderings of human-landscape relations in the context of fieldwork. And so it began—secretly at first, in our personal notebooks: the unexpected aspect of our fieldwork, a kind of an underground ethnography of archaeological fieldwork combined with the "metaphysical fieldwork" that we discuss throughout this book.

Lived and experienced landscapes are highly complex and dynamic constellations of innumerable physical, biological, historical, cultural, personal and other elements. This means that landscapes can be approached from innumerable perspectives without exhausting them; there is always more to be discovered that is relevant from one point of view or another. This book recounts how we received and responded to a "weird" landscape, working towards the broader goal of identifying some unrecognized or overlooked elements that mediate human-landscape relations beyond the predominant, objectifying descriptions of the landscape. We take it for granted that geomorphologies, vegetation and other such "objective" properties are important to past and present human-environment relations, but there are also a host of other factors, tangible and intangible, that play a role in that respect. Identifying and engaging with those "other factors" requires new tools and approaches. This is what we seek to do in this book from the angle of "weirding", which is discussed in more detail in Chapter 2.

REFERENCES

Aikio, Á. 2022. A Window into Vanishing Sámi Culture?: Visual Representations of Sáminess in the Shared Siida Exhibition by Sámi Museum Siida and Northern Lapland's Nature Centre. In *The Sámi World*, ed. S. Valkonen, Á. Aikio, S. Alakorva, and S.-M. Magga, 21–38. Abingdon: Routledge.

Davidson, P. 2005. *The Idea of North*. London: Reaktion Books.

Demiroglu, O., and C. Hall. 2020. Geobibliography and Bibliometric Networks of Polar Tourism and Climate Change Research. *Atmosphere* 11 (5): 498. https://doi.org/10.3390/atmos11050498.

Glasberg, E. 2012. *Antarctica as Cultural Critique: The Gendered Politics of Scientific Exploration and Climate Change*. New York: Palgrave Macmillan.

Herva, V.-P., and A. Lahelma. 2020. *Northern Archaeology and Cosmology: A Relational View*. Abingdon: Routledge.

Herva, V.-P., A. Varnajot, and A. Pashkevich. 2020. Bad Santa: Cultural Heritage, Mystification of the Arctic, and Tourism as an Extractive Industry. *The Polar Journal* 10 (2): 375–396. https://doi.org/10.1080/2154896X.2020.1783775.

Hovelsrud, G.K., B.P. Kaltenborn, and J. Olsen. 2020. Svalbard in Transition: Adaptation to Cross-Scale Changes in Longyearbyen. *The Polar Journal* 10 (2): 420–442. https://doi.org/10.1080/2154896x.2020.1819016.

Huggan, G., and R. Norum. 2015. The Postcolonial Arctic: Editorial. *Moving Worlds: A Journal of Transcultural Writings* 15 (2): 1–5.

Ingold, T. 2000. *The Perception of the Environment: Essays on Livelihood, Dwelling and Skill*. London: Routledge.

Ingold, T. 2010. *Being Alive: Essays on Movement, Knowledge and Description*. London: Routledge.

Lehtola, V.P. 2009. Saame-sanan juuria suomalaisessa julkisuudessa. In *Ei kiveäkään kääntämättä: juhlakirja Pentti Koivuselle*, ed. J. Ikäheimo and S. Lipponen, 37–46. Oulu: Pentti Koivusen juhlakirjatoimikunta.

Morton, T. 2013. *Hyperobjects: Philosophy After the End of the World*. Minneapolis: University of Minnesota Press.

Müller, S. 2022. From Last Chance Tourism to Gone Destinations? Future Narratives of Svalbard as a Post-Arctic Tourism Destination. Master's Thesis, Tourism Research, University of Lapland.

Paphitis, T., R. Norum, and V.-P. Herva, eds. 2021. *Time and Mind* 14 (3): Special Issue on 'Minding Arctic Fields'.

Pörtner, H.-O., D.C. Roberts, H. Adams, I. Adelekan, C. Adler, R. Adrian, P. Aldunce, E. Ali, R. Ara Begum, B. Bednar-Friedl, et al. 2022. Technical Summary. In *Climate Change 2022: Impacts, Adaptation and Vulnerability. Contribution of Working Group II to the Sixth Assessment Report of the Intergovernmental Panel on Climate Change*, ed. H.-O. Pörtner, D.C. Roberts, M. Tignor, E.S. Poloczanska, et al., 37–118. Cambridge: Cambridge University Press. https://doi.org/10.1017/9781009325844.002.

Rutty, M., S. Gössling, D. Scott, and C.M. Hall. 2015. The Global Effects and Impacts of Tourism: An Overview. In *Handbook of Tourism and Sustainability*, ed. C.M. Hall, D. Scott, and S. Gössling, 36–63. Abingdon: Routledge.

Varnajot, A., and J. Saarinen. 2021. 'After Glaciers?' Towards Post-Arctic Tourism. *Annals of Tourism Research* 91: 103205.

Open Access This chapter is licensed under the terms of the Creative Commons Attribution 4.0 International License (http://creativecommons.org/licenses/by/4.0/), which permits use, sharing, adaptation, distribution and reproduction in any medium or format, as long as you give appropriate credit to the original author(s) and the source, provide a link to the Creative Commons license and indicate if changes were made.

The images or other third party material in this chapter are included in the chapter's Creative Commons license, unless indicated otherwise in a credit line to the material. If material is not included in the chapter's Creative Commons license and your intended use is not permitted by statutory regulation or exceeds the permitted use, you will need to obtain permission directly from the copyright holder.

CHAPTER 2

Weirding and Fielding the World: Hows and Whys

2.1 Tales of Two Landscapes: Monstrous Relations and Resonances

Prior to our embarking on the Ritničohkka fieldwork, Vesa-Pekka and Oula had spent a week doing fieldwork in a place that, on the surface, is radically different from—indeed opposite to—Ritničohkka as regards the ways that humans engage with it. This other site and landscape are that of Hannukainen, a modern and now closed-down mine located some 300 kilometres south-east of Ritničohkka (Fig. 2.1). While Ritničohkka is "wilderness", insofar as such a concept makes any real sense, Hannukainen is a "destroyed" post-industrial landscape in the middle of "wilderness" where nearly five million tons of iron ore were extracted and moved around between 1975 and 1990. The task that we had set for ourselves in Hannukainen was to characterize the extraction landscape from a broadly cosmological point of view—that is, how various features of the actual landscape could be seen to resonate with broader cultural ideas about mining and the subterranean world (see Herva et al. 2024a). The Hannukainen landscape, although superficially familiar to us due to various visits there over the years, turned out to be confusing and disorienting in terms of how to approach and engage with it—how to make sense of it in "cosmological" terms. Over the week that we wandered there, it gradually started to open itself up to us and we began to process it, tentatively, in terms of monsters, decay and horror.

Fig. 2.1 View of the abandoned mining landscape at Hannukainen with the Ylläs ski resort in the background (left), and barren tundra at Ritničohkka (right)

Our processing of Hannukainen had barely started when we relocated to Ritničohkka, but we were already unconsciously "attuned" (although we didn't think about it at the time) to "weirdness" and a certain mode of seeing and thinking about the surrounding world—and to questions of what lies beneath the surface in a metaphysical sense—which we came to encounter on Ritničohkka. Despite their very different character and histories, it was easy to contrast and compare the two places, indeed to detect various affective and experiential similarities between them. For one thing, both places elicited feelings and a sense of the extraordinary. These are places that are, emotionally, partly similar and partly different. Somehow they are apart from the ordinary everyday world.

At some point, monsters and the monstrous emerged as potentially productive conceptual tools for thinking about and trying to make sense of such places, as various scholars have done in relation to the Anthropocene and to how monstrous beyond-human entities operate on our planet (Tsing et al. 2017). While Lapland is thought of and marketed as "wilderness", it has also been an arena for colonial industrial projects since early modern times, including mining (e.g. Nordin 2020), just like numerous other parts of the Arctic (e.g. Kruse 2013). Mining, in turn, is directly and indirectly a major agent and symbol of the environmental change and destruction that is integral to the Anthropocene—hence, many have found it useful to conceptualize various global processes in terms of monsters and the monstrous (e.g. Tsing et al. 2017). Moreover, the emergence and development of "civilization" since the Bronze Age is integrally, again both directly and indirectly, intertwined with extractive pursuits and industries (Herva et al. 2022). While extractive industries are central to the world as we know it today, Hannukainen and other sites of extractivism are simultaneously reminders that:

> Living in the Anthropocene implies that we are no longer at home, or that our home is no longer comfortable but filled with terrors and depths that are perhaps best captured by the metaphysics of the weird that we have uncovered in our explorations of weird fiction. Living in the Anthropocene implies not only that we are aware that the world has dimensions that exceed the grasp of our senses, but also that there are dimensions or depths to the real that exist beyond even those of science. (Tabas 2015: 36)

Ritničohkka, of course, is not isolated from the Anthropocene world and its realities. But the monstrous and largely invisible forces that are shaping the contemporary world, such as climate change, are very much present there as well. These forces merge and cooperate with others, and in some sense perhaps with more primaeval non-human and monstrous beings and forces, and the eerie "sense of presence", that are partly generated by our personal and cultural imaginaries. Partly these monstrosities are brought about by the place itself and its affordances that may guide attention and thoughts to a prehuman deep past, or fling them into the web of associations that constitutes supermodernity. In either case, both Hannukainen and Ritničohkka, whether for different or similar reasons, provoke "topographical imagination" (Henderson 2022; Joshi 2011; Tabas 2015). This is where horror fiction, such as "Lovecraft's evocations of place", comes into play as a possible source of inspiration for thinking and writing towards "disrupting our normal sense of being at home in the world" (Tabas 2015: 16)—and this is very much something that both Hannukainen and Ritničohkka provoke, with different angles and emphases on living with monsters and the monstrous in the age of the Anthropocene.

2.2 Weirding

Weirding refers to the approach to analysing and interpreting the world that recognizes weird aspects as a significant component of the world. These aspects can be, for instance, radical otherness, the unfamiliar, the defamiliarized, the uncanny, the more-than-human or non-human and the monstrous in multiple forms. Weirding draws inspiration from so-called weird fiction, pioneered by American horror writer Howard Phillips Lovecraft (1890–1937). During his lifetime and many decades after his death, Lovecraft, though highly esteemed among his peers and colleagues, was considered by most to be little more than a "pulp" writer. However, he has subsequently gained wide recognition as one of the seminal voices in expressing the anxieties and horrors of the Anthropocene, and his influence has been identified in various domains of Western popular culture (Gonzalez and Sederholm 2021). The genre was later developed by Thomas Ligotti (b. 1953), another American horror writer who came of age a generation or two after Lovecraft. Ligotti is now considered one of the founding fathers of the contemporary genre of weird (latterly "New Weird") fiction, which is contemporarily represented by authors such as Jeff VanderMeer (b. 1968) and China Miéville (b. 1972) (Iossifidis and Garforth 2022).

The analytical value of weird fiction for the study of the Anthropocene stems from the recognition that the era, characterized by a series of titanic crises—climate change being just one of them—is "weirdly weird", as Timothy Morton (2016) phrased it. Global warming itself was refashioned in the early 2000s by author and environmentalist L. Hunter Lovins as "global weirding" (see Friedman 2010), a concept that sought to describe the type of weather behaviour we haven't seen before as a way of showing where Earth is heading with continued emissions of greenhouse gases. Global weirding suggests that the world has grown highly volatile and uncertain, shaped as it is by monstrous, more-than-human forces and entities that threaten all human and non-human life on Earth. Such dark views of the planet's future are echoed in Lovecraft's fiction, which envisioned unfathomable cosmic monsters and horrors lurking just behind a thin veil of sanity that will break out to plunge the world as we know it into disarray, mayhem and ultimately an untimely end, when "the stars are right" (as Lovecraft refers to the tipping point when his fictive monsters are released).

Lovecraft's fiction underscores the insignificance of humans, showing how our lives are, ultimately, at the mercy of a weird and monstrous cosmos, one which readily resonates with the non-human powers and entities operating in (or perhaps just operating) the Anthropocene. As Gry Ulstein (2021: 21) has argued, through such literary works, "weird, weirding, or weirdness can provide a useful critical lens for studying the global environmental crisis". This "weirdly weird" world we live in "has brought nonhuman realities, human history, and geological time together in ways that deeply contradict our sense of reality" (Ulstein 2021). It is along such lines that we attempt in this book to employ weirding as a reading strategy for understanding how an Arctic landscape is situated in relation to a global socio-cultural and ecological condition.

In addition to Morton, many influential scholars and thinkers have discussed different aspects of the Anthropocene more or less explicitly in terms of the weird, horror, monsters and the monstrous within frameworks of ecocriticism and environmental humanities (e.g. Economides and Shackelford 2021; Tsing et al. 2017; Vermeulen 2020). Donna Haraway (2015, 2016), for instance, coined the term Chthulucene—which, as she underlines, was not actually inspired by the at times misogynistic and racist work of Lovecraft, but instead by a Californian spider, the *Pimoa cthulhu*. In any case, through Chthulucene, Haraway conceived the Anthropocene in terms of monstrous tentacular entanglements of humans and non-humans. Scholarly discussions of the Anthropocene and its monsters are also closely affiliated with posthumanism and object-oriented ontologies (e.g. Harman 2018), which problematize the object/subject and related binaries and decentre humans, and which, again, resonate with—and are similar to—issues featured in the framework and context of weirding. Indeed, the philosopher Graham Harman, who has championed object-oriented ontologies (e.g. 2018), employed Lovecraft's fiction in developing "weird realism" (Harman 2012; also Hegglund 2020), which also served as an inspiration for our present book.

Important and inspiring as the contributions of Haraway, Tsing, Morton, Harman and others are for thinking about the weirdness of the Anthropocene, their work tends to be theoretically quite dense—or to operate on a philosophical and abstract level—whereas our take on weirding specifically seeks to discuss the significance of the weird on a sensory and experience-based landscape phenomenological level, which also renders monsters as a useful tool (see below). In other words, we attempt to take a distinctively grounded approach to weirding: how the weird, in diverse forms and expressions, unfolds in relation to our engagement with a particular place and landscape in the "melting Arctic", and indeed on a literally and metaphorically melting planet. Our interest, then, lies with how the weird manifests itself on the level of lived experience and specifically in relation to fieldwork in an Arctic environment. We identify various forms of the weird and how we responded to them and how they affected us intellectually and emotionally.

Weird fiction—or the idea of weirding more generally—affords a certain way of seeing and sensing the landscape. It allows describing, or at least "nearing" (Olsen and Pétursdóttir 2021), and understanding them more viscerally, and ultimately linking local perceptions, feelings and experiences to more general-level matters that are integral to the Anthropocene world (see Turnbull et al. 2022 for a recent review of this territory). Weird fiction and weirding provides a perspective, a framework and vocabulary for making sense and appreciating the significance and broader implications of perceiving and relating to the environment within a broadly landscape phenomenological approach. This also involves appreciating the significance of affects, emotions, sensations and feelings in encountering and being in and with an experienced world.

Weird fiction can provide vocabulary and illustrations for making sense of different forms of weirdness. The weirdness of the Anthropocene world unfolds on an experiential and everyday level, where the abnormal and the mundane collide. Here, weird fiction affords us a particular general orientation and a way of sensing the environment, where different forms and expressions of the fieldwork context are engaged through various things and phenomena from the weirding perspective, while "on the ground". Communicating our experience of being there in the field, open to weirding, is integral to our approach in this book.

It is not easy to define the "weird". The weird is grounded in particular emotive sensations and atmospheres—"I know it when I see it", or "I know it when I feel it", as Jeff and Ann VanderMeer would have it (quoted in Brawley 2017: 89–90). The weird is related to the uncanny and the eerie, but the three are more or less different things, phenomena or experiences. A foundational work in this respect is Freud's (1919) engagement with the *unheimlich*, which is usually translated as the uncanny, though this does not quite convey the same meaning as *unheimlich*, which means "unhomely". The uncanny can be understood as minor disruptions of ordinary reality which can be strange but do not compromise the assumed basic structure of reality. The weird is often thought to be something more fundamentally strange.

Mark Fisher (2016) and others (e.g. Lockhurst 2017; Rae 2020; Heft 2021) have examined the distinctions between the weird, the eerie and the uncanny in useful ways. For Fisher (2016), "the weird is that *which does not belong*" (p. 10) and there is "[the] sense of *wrongness* to the weird" (p. 13, emphases in original). The eerie, for him, lacks the shock that is characteristic of the weird and "can give us access to the forces which govern mundane reality but which are ordinarily obscured, just as it can give us access to spaces beyond mundane reality altogether" (Fisher 2016: 13).

Very recently, "weird" has gained a political role as well. During the 2024 US presidential elections, the Democratic Vice Presidential candidate Tim Walz attached the label "weird" to his opponents Trump and Vance. It remains to be seen whether this label is effective as a political cudgel, or whether the polarizing "out-of-this-world" politics of the Anthropocene (see Latour 2018) alters the very meaning of weirdness.

In this book, we take a rather open view of what constitutes "weird"—with "weirding" serving as an umbrella term of sorts. But it can be understood as something that compromises our understanding of what reality is like or, perhaps more specifically, the established modernist view of reality. This view has for a long time depended on naturalistic, rationalist, linear, ordered, mechanistic and objectivistic ideas of the world and its workings.

Our primary aim is to identify and discuss how the weird in a broad sense—encompassing the uncanny and the eerie—is integral to the experience of the Anthropocene world, and how and why the weird matters on a general level, rather than to provide a close and detailed analysis of specific experiences. Modernity's ideas of reality have, of course, long been scrutinized and critiqued from diverse perspectives, but modernity nonetheless tends to govern "authorised" and "proper" knowledge of the world: what is "real" and what is not, what is "significant" and "important", and so forth. Weirding is about appreciating the stranger aspects of lived reality, but it is also a technique for defamiliarizing seemingly familiar things towards gaining a different perspective on the world and its workings.

The weird is something that calls such ideas into question on a fundamental level and provokes deep existentialist questions about existence, being, temporalities, spatialities and causalities. The weird tends to be associated with fundamental disruptions of reality (or what it is assumed to be like), but it is related to various other terms and concepts, such as the eerie and the uncanny. As previously mentioned, the similarities and differences between these different modalities of the extraordinary have been explored previously (e.g. Fisher 2016; Lockhurst 2017; Rae 2020; Heft 2021), and for our current purposes a rather general and tacit understanding of the weird suffices.

For us, in this book, the weird is primarily orientational and inspirational rather than a definite class of things or phenomena. In essence, we attempt to document and make sense of the "weird" Ritničohkka in relation to the contemporary Anthropocene world and human place in it as regards both visible and invisible and tangible and intangible non-human entities and powers. Faced with various forms of unsettling and weird experiences during our few days on the fjell, we started pondering on how to come to terms with those experiences and their wider significance and implications to archaeological fieldwork and conceptions of human-environment relations.

Melting glaciers and climate change can be considered as a form of "dark heritage" (Varnajot and Salim 2024; also Thomas et al. 2019) of the Anthropocene and its horrors and monsters that Lovecraft, for instance, arguably portrays in his fiction (Tabas 2015). Weirding is an exercise in trying to see and think about our lived reality differently, conceptualized as cosmic horrors and monsters, for example. Or, as Mittman puts it, "above all, the monstrous is that which creates this sense of vertigo, that which calls into question our (their, anyone's) epistemological worldview, highlights its fragmentary and inadequate nature" (Mittman 2012: 8). Weirding is a heuristic tool for discovering various elusive but nonetheless real, more-than-human entities and forces that operate in the world. The experienced reality in this view is richer in contents and dynamics than the predominant Western modernist, materialistic and mechanistic thinking proposes.

This is a particularly suitable view on the (sub-)Arctic not only because processes such as global climate and environmental change are particularly prominent there, but also because northern ways of life and thinking are characterized by relational ontologies and epistemologies (Herva and Lahelma 2020). Weirding, as it features in the relevant academic literature, provides a view on the high modern Anthropocene world and how it came about. It is a "lens" rather than a classificatory system in which some things or places are "weird" or "monstrous" and others not, although certain things and places may resonate more with weirding perspectives than others.

Both relational thinking and weirding seek to transcend the modernist boundaries between, for instance, visible/invisible realities, object/subject, material/spiritual and presence/absence. Our experience of places and being in the world are affected and shaped by diverse more-than-human beings and powers that are not necessarily readily perceivable, nor purely subjective or objective or natural or cultural, and thus cannot be properly grasped with modernist terms.

Relationality and weirding are tools for trying to understand the deeply dynamic and networked lived reality, in particular its more-than-human and beyond-the-rational dimensions. In this view, monsters and spirits are partly about the world "out there" and partly in the mind (as indeed befits the ambiguous nature of monsters), but not reducible to either. The landscapes of Ritničohkka generated feelings and sensations of invisible and hidden presences which were integral to our experience of the place but also raised questions of how to do and write about archaeological fieldwork and archaeology in general (see Olsen and Pétursdóttir 2021).

Many of these potentially significant aspects of land cannot be "objectively" recorded, which highlights the significance of the "human body as a medium of exploration" (Tuominen and Marila 2021; Seamon 2018) or as a "conduit" between different dimensions of reality, making autoethnography a useful approach. The body is an instrument of knowing in the form of emotions, for instance, which provide information on certain aspects of human-environment interaction, and this becomes particularly acute when working in unfamiliar landscapes.

Such knowledge is situational and based on the simultaneous awareness of what happens in the environment and in us (Bird-David 1999; Ingold 2000). It may involve experiences of a heightened sense of awareness, with one becoming more attuned to one's surroundings, which potentially provides insights into affordances and affects engendered by the land itself, such as a sense of wonder. At the same time, it also constitutes a transformative experience for us as "fielders", challenging our understanding of the world and our place in it. This entails encounters between all kinds of entities, both "real" and "imagined"—or rather, relationally constituted—which can be expected to occur especially in unfamiliar landscapes that can generate dream-like perceptions of the world, altered states of consciousness and an otherworldly sense of place.

Archaeology is a weird business to begin with. Posthumanism and object-oriented ontologies as well as various aspects of the Anthropocene have been subject to substantial interest in more theoretically inclined circles of archaeology and heritage studies throughout the 2000s (the literature on theoretical perspectives to Anthropocene archaeology and heritage is rather extensive, see, e.g. Olsen and Pétursdóttir 2014, 2021; Hodder 2012; Edgeworth et al. 2014; Lane 2015; Edgeworth 2021; Pétursdóttir 2017; Bangstad and Pétursdóttir 2021; Harrison 2015; Harrison and Sterling 2020). While the current thinking and research in archaeology does not directly or explicitly engage with weirding, this literature does feature weird entanglements between people, things and the world, albeit discussed in different terms.

There are also other angles on the entanglements of archaeology and the weird, as exemplified by archaeology's intertwinement with supernatural and otherworldly matters in popular culture and "fringe" or "alternative" archaeologies (e.g. Card 2018). The analytical or inspirational value of ghosts, hauntings and the uncanny—in part inspired by fiction—has also been considered in archaeology (e.g. Clarke 2007; Moshenska 2006, 2012; Herva 2014; Herva and Matila 2024, forthcoming).

While archaeology is, then, quite naturally a home of the weird, it still awaits methodological weirding. As far as we can tell, there is only one PhD thesis that has specifically engaged with archaeology in terms of weirding, namely Kerry Dodd's (2020) dissertation, *The Archaeological Weird: Excavating the Nonhuman*, which, however, is somewhat differently oriented than this book. Nonetheless, these examples have laid the groundwork for the archaeological methodology of weirding and for our metaphysical journey.

2.3 Monsters and the Monstrous

The weird may or may not necessarily be associated with monsters and the monstrous, but for Lovecraft's fiction, for instance, unfathomable and indescribable monsters—such as those by the name of Cthulhu, Nyarlathotep and Shub-Niggurath—are critically important devices or expressions of the weird. They are entities that are fundamentally about wrongness in the prevailing human perspective of reality. Monsters and the monstrous, as we see them, make the weird something concrete and experiential—pointing to things that are, in one sense or another, outside of or extending beyond the ordinary world. Monsters are about cognitive dissonance (Wengrow 2014; Martín Porras 2022). They are jars and vaults in and of reality. Hence, for our purposes, the monstrous is closely related to the weird. Monsters (or the weird) are not necessarily about something supernatural, except in the technical sense that both monsters and the weird violate the categories of modernist ordinary reality.

We humans today live with manifold monsters, just like people did in the early civilizations of the Near East and Egypt. There appears to be some meaningful connection between complex state societies and monsters, whether ancient or super-modern (see Wengrow 2014 on the birth of monsters). When we talk about monsters in this book, we primarily have in mind modern monsters, real and fictitious. Actually, monsters *are* real. But in what sense, in what way? It is not so easy to define monsters or say what they are, which is why this question must be considered here at some length.

The contemporary age of the Anthropocene is very much a time of monsters, real and metaphorical, and we are surrounded by all kinds of monsters and the monstrous. While the kinds of things that come to mind when thinking of monsters—such as dragons and griffins—may not exist as real-world autonomous embodied beings, monsters nonetheless persist within and beyond popular culture and modern folklore. Sometimes monsters have distinctive real-world effects, as is readily demonstrated by the so-called "Slender Man stabbing" in 2014, in which two 12-year-old Wisconsin girls repeatedly assaulted a classmate 19 times with a knife, ostensibly to appease the fictional horror character Slender Man.

While some monsters have a conventionalized form and shape with determinable boundaries to their being—take dragons and griffins as an example again—monsters, and monsters of the Anthropocene in particular, are elusive and in many ways indeterminate entities, which is also why monsters and the monstrous are useful for our purposes in this book. Such monsters are, as previously mentioned, exemplified by H. P. Lovecraft's cosmic monstrous entities. As many scholars have recently noted (see, e.g. Tsing et al. 2017; Ulstein 2021), various real-world phenomena in the Anthropocene can be regarded as monsters or monstrous and, in theoretical terms, as similar to the monsters conceived in weird fiction—including climate change, pollution, environmental degradation and many other processes and phenomena of our world, which endanger and wipe out the lives of humans and non-human beings and threaten all life on

Earth. These global horrors, or monsters, are very real and simultaneously indefinite—they are visible and invisible, tangible and intangible, with boundaries, natures, spatialities and temporalities that are radically different from those of humans and other familiar lifeforms (e.g. Holloway 2017). Monsters, then, are about the unknown and radical otherness, among other things.

Moreover, for Lovecraft, there is a much more vast and chaotic "real" reality of monstrous entities beneath the thin veil that people ordinarily perceive in their daily life, which they mistake for reality. Monsters are a tool for describing, thinking, analysing and interpreting the weirdness of the world or of being in and experiencing the world—much as it presented itself to us in the context of Ritničohkka. Monsters come in many forms, and they are, by nature, difficult to pin down in definitive terms. In visual and material culture, monsters often feature as composite or hybrid—and thus unnatural—beings. Monsters are liminal beings in that they are otherworldly in some respects, yet this-worldly in others.

Monsters have served various functions in different cultural contexts, including as portents and warnings. Indeed, etymologically, the word monster is related to the concept of *revealing* something (see Mittman 2012). Some monsters have a tangible body and others—such as Arabian wind demons—have a more intangible or ethereal existence, although they can take (or at least be represented in) a conventionalized form. Lovecraft's cosmic monsters are beyond human description and comprehension and can only be vaguely alluded to with words or images (see Harman 2012; see also Fisher 2016).

Monsters and the monstrous—and monsters of the Anthropocene in particular, in both the Lovecraftian and theoretical sense—can be considered to have similar qualities as the weird; indeed, they are illustrative manifestations of the weird. Monsters are not only simultaneously subjective and objective and this-worldly and otherworldly, but they challenge linear space–time and presence/absence relations and thus extend into and operate partly beyond the boundaries of the ordinary or rationalistically conceived reality. Monsters have visible and invisible dimensions to them, and there is uncertainty about their nature and being as entities: they are ambiguous and ambivalent, provoking cognitive dissonance. In other words, there is a distinctive "more-ness" or "excess" to them in comparison with "ordinary" or "natural" entities (see further, e.g. Holloway 2017; Mittman 2012; Musharbash 2014). All this makes monsters "good to think with" (Mittman 2012: 8) when exploring the weirder and unfamiliar aspects of lived reality. Ureta and Flores (2018) have discussed active tailing ponds, or mining dumps, as behaving like dragons and tricksters, which illustrates how monsters and the monstrous conjoin with the modern industrial landscape.

The boundary is not clear between monsters and other more or less person-like, non-human beings that flourish in, for instance, the folklore of Finns, the Sámi and Scandinavians. But as with what qualifies as weird, a general understanding of what monsters are, as delineated above, is quite sufficient for our purposes here. In any case, weird monsters in weird fiction go together with "the weird reality of the Anthropocene" (Bradshaw 2020: 4), which "asks us [...] to acknowledge the failures of our systems of categorization" (Mittman 2012: 8).

2.4 Bodies on the Move

The weird is an emotional response to sensory perception, which readily highlights the significance of the body and the bodily in encounters with the weird. Thus, just as weirding can be understood as associated with literary geographies, the weird—as pursued in relation to extractive industries, for instance—is also related to geographies of emotions (e.g. Wright 2012; Ey and Sherval 2016; Ey et al. 2017). While mines and mining landscapes may initially appear quite unrelated to the landscapes of Ritničohkka, they have many similarities that are relevant within the perspective of weirding, with landscapes of extraction featuring myriad "existentially disturbing" aspects (Herva et al. 2024a, 2024b; see also Herva et al. 2022).

While we were conducting our survey on Ritničohkka, we became keenly aware of our bodily responses to the landscape, initially due to the sheer difficulty of navigating it. This resulted largely from the rockiness of the terrain, combined with the steep slope where the glacier had been. We had to move carefully, while simultaneously keeping our eyes open for any finds of interest that may have emerged from the ice. Indeed, we noticed our different approaches to moving in the landscape as soon as we embarked on the survey, with some of us moving gracefully, almost gliding from the top of one boulder to another, while others crawled more clumsily among the rocks.

The difficult terrain did not merely affect our movement in a mechanical sense, but also heightened our awareness of various features of the landscape, thereby making us particularly responsive—physically, cognitively and emotionally—to our surroundings, and consequently also perceptive of and receptive to our feelings and sensations, which are at the heart of weirding. This created a foundation for recognizing various resonances and correspondences between our perceptions, mental states, and diverse features and aspects of the landscape, especially those that struck us as peculiar or intriguing in one form or another. Later on, the changing weather and closer engagement with the cabin that served as our base camp re-calibrated our perception and sensations of Ritničohkka, as we will detail and discuss throughout the book.

Overall, the way we felt about the Ritničohkka landscape was that we were, in a very distinctive manner, being and working together with it in a reciprocal relationship in which the landscape facilitated or disabled particular ways of being, moving and doing. Having studied northern relational ways of knowing the world for many years, we found that our sense of Ritničohkka emphasized, in a very concrete manner, the significance of a highly situational and contextual mode of perceiving and knowing the landscape, anchored in constantly changing bodily-cognitive-emotional connections and correspondences between ourselves and our immediate and overall surroundings.

In theoretical terms, such situational knowing is affiliated with relational epistemologies and ontologies, animism (old and new) and shamanism. Shamanism, in particular, is related to ideas of a layered cosmos, with some of its invisible dimensions accessible through, for example, altered states of consciousness, which Ritničohkka induced, in however mild forms, in various instances. Historically and culturally, these relationalities are integral to northern ways of knowing and being in the world (e.g. Ingold 2000, 2015; Herva and Lahelma 2020).

2.5 Making Sense of the (Anthropocene) World Through Fieldwork

The conventional premise of archaeological and anthropological fieldwork is to collect data. But fieldwork has many other aspects to it as well. This book discusses and emphasizes the sensory, affective and transformative dimensions of fieldwork. We are particularly interested in how encounters with a landscape (in our case a high northern landscape) affect people and their more general perceptions and ideas of specific places, and ultimately their relationships with and understanding of the world and its workings. In other words, we regard fieldwork as a particular dynamic form of human-environment relations.

All five of us authors have done fieldwork in Lapland, (sub-)Arctic Europe and elsewhere in high-latitude environments. For the most experienced of us, this work has continued for more than 20 years. Our earlier fieldwork, archaeological and anthropological, has been highly diverse in scope. We have done conventional archaeological surveys and excavations focused on prehistoric sites as well as on historical and contemporary sites. Recent themes have alluded to, for instance, reindeer herding, the Second World War and Lapland tourism. We have done fieldwork in traditional scholarly frameworks but also in collaboration with various stakeholders in the context of community approaches to heritage. All this generated, some years ago, an interest in fieldwork as a practice and a mode of perceiving and engaging with high northern worlds (see Paphitis et al. 2021), which this book seeks to develop further.

The approach that we take could be described as an ethnographic account of archaeological fieldwork with a focus on human-landscape relations, conducted by four archaeologists with theoretical and anthropological inclinations and one cultural anthropologist (cf. Hamilakis and Anagnostopoulos 2009; Hamilakis 2011). Our approach is broadly autoethnographic and auto-archaeological in the sense that besides doing the more conventional side of our fieldwork we were attentive to how we related to our environment with its diverse experiential and affective aspects—that is, how we "connected" with Ritničohkka, both mentally and materially, and how Ritničohkka "connected" with us. Or, to put it in slightly more specific terms, we tried to engage with the relational and reciprocal or dialogic dimensions of our engagement with Ritničohkka. Relational thinking comes in many specific forms and under many banners, such as indigenous perspectivism (e.g. Viveiros de Castro 1998) and posthumanism, which is distinctively rooted in contemporary Western ideas (e.g. Harman 2018).

This book, then, is about how to make sense of weird landscapes—or the challenges of such a pursuit—and human entanglements with such landscapes. Our take is theoretical in the general sense that we are interested in frameworks of thinking and how to connect and engage especially with some weirder aspects of landscapes and their evocative and affective qualities, conceived as relationally constituted and known. Rather than looking for definitive answers, however, we discuss perspectives anchored in our experiences from a broadly autoethnographic perspective and in an explorative manner.

Fieldwork, in archaeology and other disciplines, has long been a domain of methods rather than theory. There are, however, significant exceptions, such as Ian Hodder's (2014) fieldwork at Çatalhöyük. Recently, there has been an increasing interest in various fields in aspects of fieldwork other than pure data collection and documentation. Couper, for instance, writes that "Physical geography fieldwork entails more than is written into the publication of results. Being-in-the-field is a multidimensional, cognitive, socio-cultural, corporeal, affective experience that is filtered in order to arrive at the propositional knowledge that counts as science" (Couper 2023: 430).

This book takes a similar perspective on archaeological fieldwork through the case of our experiences in the remotest corner of Finnish Lapland. We were also inspired by Marr's and her co-authors' recent article subtitled "reflections of more-than-human field/work encounters", which quite aptly characterizes our take in this book as well (Marr et al. 2022). It is also increasingly recognized that novel experimental approaches and conceptual frameworks are needed to describe landscapes and fieldwork. Thus, for instance, Lantto has experimented with "incorrectly exposed photography as an alternative way of understanding geomorphological research" (Lantto 2020).

It is along similar lines that we also attempt to "see" a high northern landscape differently and to narrate our fieldwork differently (cf. Olsen and Pétursdóttir 2021), which makes this book purposefully experimental in character. This entails employing conceptual tools or ideas derived from the studies of haunted and haunting landscapes (Matila 2020; Paphitis 2020), and especially from monster studies and, of course, the aforedetailed "weirding" approaches to the world (Musharbash 2014; Tabas 2015; Holloway 2017; Turnbull et al. 2022). These ideas and perspectives are employed towards appreciating the significance of various non-linear and relational dimensions of human-land entanglements.

The snowfield on Ritničohkka was not only the primary target of our fieldwork but also the primary anchor point for the more existentialist questions that our experience and engagement with our research site sparked during the few days that we spent there (Fig. 2.2). Just a few years ago, we did not even know that there was a glacier-like formation in Finland, which goes some way towards demonstrating the degree of otherness that we expected to encounter (though some of us had been on "real" glaciers elsewhere). Melting glaciers—glacier extinction—in polar regions readily exemplify the global climate change that is a characteristic of the Anthropocene. Yet, for most Westerners climate change still tends to be present in a distanced or abstracted manner, in the form of nature documentaries, news reports and policy statements.

The concept of the Anthropocene has been the subject of much debate and discussion over the last twenty or so years, and we use it simply as a context. Opinions vary as to when the Anthropocene as a putative geological epoch began, with arguments ranging from the Early Neolithic to the mid-twentieth century to not-just-yet, but the basic idea is that humans have come to have a significant, geological, ecological and climate impact on the planet. Today, we live with a global climate crisis and environmental anxiety, which is symbolized by, for instance, melting glaciers. This brings a direct connection, lurking in the background, to our fieldwork in Ritničohkka. Geologists have recently debated whether the Anthropocene is really a "geological epoch" or an event (Gibbard et al. 2022; Head et al. 2023; IUGS 2024), but this does not really undermine the significance of the Anthropocene as a socio-cultural phenomenon in the humanistic and social sciences (Brauch 2021; Wallenhorst 2023); the Anthropocene is here, whether we want it or not.

Fig. 2.2 View of the infinite Arctic tundra from the summit of Ritničohkka

Moreover, the Anthropocene—or the contemporary world specifically—is increasingly described in terms of monsters, which in turn readily connects it to Weird Fiction and provides tools for our attempts to grasp how we perceived and related to the environment on Ritničohkka. Finally, monsters offer a bridge or a common ground for dialoguing, or at least making connections, between certain theoretical ideas and local relational northern cosmologies and modes of engaging with the environment. Monsters of the Anthropocene are very different entities from the sentient, person-like beings described in Finnish, Scandinavian and Sámi folklore and related sources, for instance, but they also have some common characteristics and qualities in that they transcend various boundaries of reality as conceived in modernist linear, mechanical and dualist terms.

On Ritničohkka, we were presented with a haunting presence in the broader setting of an "exotic" Lapland. Our encounters with and experiences of this place and land in the context of our fieldwork pushed us to think about and approach climate change and related processes in a new way, especially in regard to the more specific question of how to characterize places that in many ways look and feel not just unfamiliar but weird, in comparison, that is, with our everyday landscapes and how we relate to them. This book, as stated, approaches the Ritničohkka landscape from a broadly landscape phenomenological perspective (e.g. Tilley 1999, 2008; Johnson 2012), which entails wandering in the landscape with attentiveness to its various elements, features and qualities, as well as to the relationships between its various "natural" and "cultural" constituents, from topographies and landforms to colours and textures, feelings and atmospheres.

2.6 Fielding in Practice

Our approach to fieldwork has evolved through continued and recurrent cooperation over time. We have come to carry out our fieldwork in practically non-hierarchical, small teams composed of people who know each other quite intimately and have been working together for years. Familiarity with each other's peculiarities, strengths and weaknesses promotes tolerance and—most importantly—trust in each other. Fun and enjoyment are key factors when fielding together.

These last points have become exceedingly important over time. Without mutual trust and an open-minded, playful approach to work, we never would have made most of the observations presented in this book. The significance of playfulness, in particular, cannot be stressed enough in this context, as the modern neoliberal work life and academic circles typically offer far too few possibilities for open-ended experimentation and play, which are vital elements in any creative process.

Also, close cooperation and co-production with people from different backgrounds has helped to broaden the perspectives and scope of studies, bringing multivocality to the interpretations. These collaborations have included indigenous people (specifically, in Northern Finland, the Sámi) and various local and other stakeholders, from archaeology and history buffs to memorabilia collectors and metal detectorists. With local Sámi contacts, for nearly 20 years or so, we have carried out mutually beneficial collaborative work, typically initiated by local needs and interests. This collaboration has now become a working template for various field studies in the North (Herva et al. 2024c).

A collaborative co-production of knowledge has turned out to have much higher social and communal relevance and tangible benefits for the local partners in various ways than projects born solely from academic and scientific interests and perspectives do. This—sometimes unexpected and unanticipated (see Herva and Seitsonen 2020)—social importance and co-producing can challenge the hierarchical, top-down views from the proverbial ivory tower, which are still widespread in many academic and heritage management projects coordinated from the dominant centres. Local collaboration has a better chance of actually initiating and making some meaningful change on a local level, no matter how small.

Local interest was also vital for our work in the Kilpisjärvi region. Although we had been carrying out conflict archaeological work in the area since 2015 (e.g. Seitsonen et al. 2021; Stichelbaut et al. 2021), the work described in this book was sparked by a call from the head of the local Sámi herders, reindeer chief Juha Tornensis, who contacted the University of Oulu in 2018. He asked if we could assist in recording the local Sámi heritage from their own perspective, in order to get some official recognition for their long-enduring (albeit seasonal) presence and land use history in the area, which is often neglected in Finnish narratives that underline Kilpisjärvi's existence as a very recent Finnish village (Herva et al. 2024c). Since then, we have carried out archaeological and heritage work in the area and, strolling across the hills and tundra with Juha, have been guided to see the landscape through Sámi eyes. Our work has, for instance, established the antiquity of Sámi reindeer pastoralist settlement in the area, dating back over a millennium (Seitsonen 2020; Seitsonen and Viljanmaa 2021).

In general, we plan our field trips on the loosest possible basis, allowing time and space for improvisation and for the unexpected twists and turns typically encountered in the northern "wild". Northern understandings of time and tempo can be quite different from what you encounter in the often-hectic, more efficiency-driven mid-northern latitudes of Euro-America. As the most obvious and easily understandable hindrance for planning specific hourly, or even daily, timetables, the sometimes extreme changes in weather can completely alter fieldwork prospects. In northern environments, hastiness can easily turn fatal, or at least dangerous. The locals also usually have a very different approach to timetables than people living in the dominant administrative regions, and as we often cooperate with reindeer herders, for example, we must acknowledge their sense of time and priorities. In the herders' case, these are dominated by the needs and the comings and goings of their reindeer, so we cannot plan too specifically in advance when cooperation will occur. This lack of fixed plans and timing can sometimes be a source of great frustration for people used to a more scheduled way of life, but it is, in fact, very Northern in its essence.

On Ritničohkka, we found ourselves engaged with this land in a kind of a *dérive* mode (Debord 1956). *Dérive*, or "drifting", can be understood as open-ended and unplanned wandering in a landscape, which potentially affords detecting connections and associations between diverse material landscape elements and cultural ideas (Herva and Rapakko 2023). Inspiring as this unplanned drifting aspect can be, it does not pretend to mimic how, say, Sámi reindeer herders or past inhabitants relate to this landscape. What it does, however, is to cultivate awareness of and attentiveness to different aspects of the dynamics of human-landscape relations. In other words, it affords perceiving the environment differently from the accustomed ways, providing unexpected connections and associations between things. This serves, first and foremost, heuristic purposes for seeing and thinking about immensely rich and multidimensional human-landscape relations. Fielding in the *dérive* way is not about ascertaining specific meanings of landscapes, past or present, but about tracing and expanding interpretive horizons and possibilities.

After carrying out our field forays in this semi-improvised, open-ended manner in Sápmi/Lapland over the course of a decade (see Seitsonen 2021: 24), we have come to notice that other people elsewhere have been theorizing about similar issues regarding fieldwork. Most notably, recent anarchist approaches to archaeological practice have underlined the need to carry out fieldwork in a way that challenges the traditional hierarchical and authoritarian neoliberal models of conducting archaeological work (Angelbeck et al. 2018; Black Trowel Collective 2016). Anarchist archaeologists often embrace the principles of collective action, horizontal decision-making, mutual aid and support, as well as close collaboration with or participation by local or descendant communities in the work. Anarchist community archaeology aims to drive social change and also seeks to decolonize archaeology, which famously has its own long-standing and heavy colonial burden (Black Trowel Collective 2016). The decolonization of archaeology and the past is a prevalent issue in Sápmi (e.g. Hood 2015), and surprisingly even some archaeologists who have worked for decades in the area can be rather blind to the continuing settler colonial issues and mindsets and how these connect with archaeological research and interpretations (see Nylander 2023).

Inspired by different branches of relational thinking, we seek to operationalize them on a general level. Different ways of relational thinking often aspire to go beyond mechanistic and dualistic visions of the world and to conceive dynamic two-way relations between humans and all kinds of non-human entities, which affect and constitute each other. In brief, we tried to be sensitive to how these theoretical ideas (and related ideas examined in Weird Fiction) resonated with our perceptions and experiences of Ritničohkka. We were fielding the field, but simultaneously we were also, to use Paphitis's (2021) clever turn of phrase, "fielding the mind" in relation to an Arctic landscape. Overall, we tried to get a sense of relational knowing as Bird-David (1999) considers it: being aware and conscious of how certain engagements with non-human entities affect the different parties in a specific context of interaction. Importantly, as Ingold (1999) points out, relational knowing is not something limited to indigenous cultures but is still here with us. It has not disappeared but has instead lost its authority to the institutions of modern society.

Jokingly, among ourselves, we referred to our fieldwork as an "expedition". The word, of course, resonates with ideas of polar exploration, such as Amundsen venturing to the North and South Pole, and with a host of related themes of unknown and unexplored lands, along with various fictional accounts of such places. Remote lands clearly still exert a pull on us, despite technological and other developments that have made remote corners of the globe much more easily—and safely—accessible than during Amundsen's time. We have accurate maps, Global Positioning System, hi-tech clothes, and many other things and systems in place that reduce the risk of dying in or being seriously injured in remote lands and extreme conditions. At the same time, however, this technology not only runs the risk of creating a false sense of safety, but also flattens and homogenizes reality. Yet, the actual "being there" presents a very different story of the land as an experienced reality and its multiplicities. In what follows, we examine and reflect on different modalities of encountering and engaging with this landscape and on how to describe and understand its dynamics, multiplicities and dissonances with regard to human-environment interaction, drawing from the so-called weirding approaches and founded on our experiences on Ritničohkka.

REFERENCES

Angelbeck, B., L. Borck, and M. Sanger. 2018. Anarchist Theory and Archaeology. In *Encyclopedia of Global Archaeology*. Cham: Springer. https://doi.org/10.1007/978-3-319-51726-1_2627-1.

Bangstad, T.R., and Þ. Pétursdóttir, eds. 2021. *Heritage Ecologies*. Abingdon: Routledge.

Bird-David, N. 1999. "Animism" Revisited: Personhood, Environment, and Relational Epistemology (with Comments). *Current Anthropology* 40 (Supplement): 67–91.

Black Trowel Collective. 2016. Foundations for an Anarchist Archaeology: A Community Manifesto. Savage Minds, October 16. https://savageminds.org/2016/10/31/foundations-of-an-anarchist-archaeology-a-community-manifesto/.

Bradshaw, A. 2020. Accessing Microbial Lifeworlds: Weird Entanglements and Strange Symbionts. *PULSE: The Journal of Science and Culture* 7 [online]: 1–21.

Brauch, H.G. 2021. The Anthropocene Concept in the Natural and Social Sciences, the Humanities and Law: A Bibliometric Analysis and a Qualitative Interpretation (2000–2020). In *Paul J. Crutzen and the Anthropocene: A New Epoch in Earth's History. The Anthropocene: Politik—Economics—Society—Science*, vol. 1, ed. S. Benner, G. Lax, P.J. Crutzen, U. Pöschl, J. Lelieveld, and H.G. Brauch. Cham: Springer. https://doi.org/10.1007/978-3-030-82202-6_22.

Brawley, C. 2017. "The Icy Bleakness of Things": The Aesthetics of Decay in Thomas Ligotti's "The Bungalow House." *Studies in the Fantastic* 4: 82–100.

Card, J.J. 2018. *Spooky Archaeology. Myth and the Science of the Past*. Albuquerque: University of New Mexico Press.

Clarke, P.A. 2007. Indigenous Spirit and Ghost Folklore of "Settled" Australia. *Folklore* 118 (2): 141–161. http://www.jstor.org/stable/30035418.

Couper, P.R. 2023. Interpretive Field Geomorphology as Cognitive, Social, Embodied and Affective Epistemic Practice. *Canadian Geographies/ Géographies canadiennes* 67 (3): 430–441. https://doi.org/10.1111/cag.12821.

Debord, G. 1956. Theory of the dérive. *Les Lèvres Nues* 9. Available at: https://www.larevuedesressources.org/theorie-de-la-derive,038.html. Accessed 25 June 2023.

Dodd, K. 2020. The Archaeological Weird: Excavating the Nonhuman. PhD thesis, Lancaster University.

Economides, L., and L. Shackelford. 2021. Introduction: Weird Ecology: VanderMeer's Anthropocene Fiction. In *Surreal Entanglements: Essays on Jeff VanderMeer's Fiction*, 1–26. Abingdon: Routledge.

Edgeworth, M. 2021. Transgressing Time: Archaeological Evidence in/of the Anthropocene. *Annual Review of Anthropology* 50: 93–108. https://doi.org/10.1146/annurev-anthro-101819-110118.

Edgeworth, M., J. Benjamin, B. Clarke, Z. Crossland, E. Domanska, A.C. Gorman, P. Graves-Brown, E.C. Harris, M.J. Hudson, J.M. Kelly, V.J. Paz, M.A. Salerno, C. Witmore, and A. Zarankin. 2014. Archaeology of the Anthropocene. *Journal of Contemporary Archaeology* 1 (1): 73–132. https://doi.org/10.1558/jca.v1.i1.73.

Ey, M., and M. Sherval. 2016. Exploring the Minescape: Engaging with the Complexity of the Extractive Sector. *Area* 48 (2): 176–182.

Ey, M., M. Sherval, and P. Hodge. 2017. Value, Identity and Place: Unearthing the Emotional Geographies of the Extractive Sector. *The Australian Geographer* 48 (2): 153–168.

Fisher, M. 2016. *The Weird and the Eerie*. London: Repeater.

Freud, S. 1919. Das Unheimliche. *Imago* 5 (6): 297–324. https://archive.org/details/Imago-ZeitschriftFuumlrAnwendungDerPsychoanalyseAufDie_464/page/n1/mode/2up.

Friedman, T. 2010. Global Weirding Is Here. *New York Times*, February 17. https://www.nytimes.com/2010/02/17/opinion/17friedman.html. Accessed November 13, 2023.

Gibbard, P., M. Walker, A. Bauer, M. Edgeworth, L. Edwards, E. Ellis, S. Finney, J.L. Gill, M. Maslin, D. Merritts, and W. Ruddiman. 2022. The Anthropocene as an Event, Not an Epoch. *Journal of Quaternary Science* 37: 395–399. https://doi.org/10.1002/jqs.3416.

Gonzalez, A.A., and C.H. Sederholm, eds. 2021. *Lovecraft in the 21st Century: Dead, but Still Dreaming*. Abingdon: Routledge.

Hamilakis, Y. 2011. Archaeological Ethnography: A Multitemporal Meeting Ground for Archaeology and Anthropology. *Annual Review of Anthropology* 40 (1): 399–414.

Hamilakis, Y., and A. Anagnostopoulos. 2009. What Is Archaeological Ethnography? *Public Archaeology* 8 (2–3): 65–87. https://doi.org/10.1179/175355309X457150.

Haraway, D. 2015. Anthropocene, Capitalocene, Plantationocene, Chthulucene: Making Kin. *Environmental Humanities* 6 (1): 159–165.

Haraway, D. 2016. *Staying with the Trouble: Making Kin in the Chthulucene*. Durham: Duke University Press.

Harman, G. 2012. *Weird Realism: Lovecraft and Philosophy*. Winchester: Zero Books.

Harman, G. 2018. *Object-Oriented Ontology: A New Theory of Everything*. London: Pelican UK.

Harrison, R. 2015. Beyond "Natural" and "Cultural" Heritage: Toward an Ontological Politics of Heritage in the Age of Anthropocene. *Heritage & Society* 8 (1): 24–42. https://doi.org/10.1179/2159032X15Z.00000000036.

Harrison, R., and C. Sterling. 2020. *Deterritorializing the Future: Heritage in, of and After the Anthropocene*. London: Open Humanities Press.

Head, M.J., J.A. Zalasiewicz, C.N. Waters, S.D. Turner, M. Williams, A.D. Barnosky, W. Steffen, M. Wagreich, P.K. Haff, J. Syvitski, R. Leinfelder, F.M. McCarthy, N.L. Rose, S.L. Wing, Z. An, A. Cearreta, A.B. Cundy, I.J. Fairchild, Y. Han, J.A.I.D. Sul, C. Jeandel, J. McNeill, and C.P. Summerhayes. 2023. The Anthropocene Is a Prospective Epoch/Series, Not a Geological Event. *Episodes* 46: 229–238. https://doi.org/10.18814/epiiugs/2022/022025.

Heft, P. 2021. Betwixt and Between: Zones as Liminal and Deterritorialized Spaces. *Pulse: The Journal of Science and Culture* 8 (2): 1–20.

Hegglund, J. 2020. Unnatural Narratology and Weird Realism in Jeff VanderMeer's Annihilation. In *Environment and Narrative: New Directions in Econarratology*, ed. E. James and E. Morel, 27–44. Ohio: Ohio State University Press.

Henderson, D. 2022. A Tale of Two Providences: Topographical Realism in 'The Haunter of the Dark.' *Lovecraft Annual* 16: 3–22. https://www.jstor.org/stable/27204672.

Herva, V.-P. 2014. Haunting Heritage in an Enchanted Land: Magic, Materiality and Second World War German Material Heritage in Finnish Lapland. *Journal of Contemporary Archaeology* 1 (2): 297–321.

Herva, V.-P., and A. Lahelma. 2020. *Northern Archaeology and Cosmology: A Relational View*. Abingdon: Routledge.

Herva, V.-P., and T. Matila. 2024, in press. Excavating the Spirit(s) of a Haunted City: Presences, Absences and Ghosts in Oulu, Finland. In *Existences Beyond Existences: Archaeology, Affect and Co-presences*, ed. J.R. Pellini and M. Bezerra. Madrid: JAS Arqueología Editorial.

Herva, V.-P., and J. Rapakko. 2023. Insides, Outsides and the Labyrinth: Knossos, Palatial Space and Environmental Perception in Minoan Crete. *Journal of Social Archaeology* 23 (3): 264–285. https://doi.org/10.1177/14696053231186771.

Herva, V.-P., and O. Seitsonen. 2020. The Haunting and Blessing of Kankiniemi: Coping with the Ghosts of the Second World War in Northernmost Finland. In *Entangled Beliefs and Rituals*, ed. T. Äikäs and S. Lipkin, 225–235. Helsinki: The Archaeological Society of Finland.

Herva, V.-P., T. Komu, and T. Paphitis. 2022. Extraordinary Underground: Fear, Fantasy and Future Extraction. In *Resource Extraction and Arctic Communities: The New Extractivist Paradigm*, ed. S. Sörlin, 166–182. Cambridge: Cambridge University Press.

Herva, V.P., G. Moshenska, T. Paphitis, T. Äikäs, I. Banks, R. Nurmi, O. Seitsonen, and S. Thomas. 2024a. Weird Quantities: Characterising Monstrous Landscapes of Extraction in the Anthropocene. *Time and Mind*: 1–24 (online first). https://doi.org/10.1080/1751696X.2024.2353748.

Herva, V.-P., O. Seitsonen, T. Paphitis, T. Komu, G. Moshenska, and R. Nurmi. 2024b. Spiralling into a Labyrinth of Cultural Fantasies and Extractivism: Treasures, Extraordinary Undergrounds, and the 'Temple of Lemminkäinen' (Sipoo, Finland). In *Connecting with Ambivalent Heritage: Creative Uses of Postindustrial Spaces*, ed. T. Äikäs and T. Matila, 131–154. London: Bloomsbury.

Herva, V.-P., O. Seitsonen, I. Banks, G. Moshenska, and T. Paphitis. 2024c. Folk Magic and the Haunting of the Second World War in Finnish Lapland. *Cambridge Archaeological Journal*: 1–19 (online first). https://doi.org/10.1017/S0959774323000495.

Hodder, I. 2012. *Entangled: An Archaeology of the Relationships Between Humans and Things*. Chichester: Wiley-Blackwell.

Hodder, I. 2014. Çatalhöyük: The Leopard Changes Its Spots: A Summary of Recent Work. *Anatolian Studies* 64: 1–22.

Holloway, J. 2017. On the Spaces and Movements of Monsters: The Itinerant Crossings of Gef the Talking Mongoose. *Cultural Geographies* 24 (1): 21–41.

Hood, B.C. 2015. Framing Sami Entanglement in Early Modern Colonial Processes: Ethnohistorical and Archaeological Perspectives from Interior North Norway. *Arctic Anthropology* 52 (2): 37–56. https://doi.org/10.3368/aa.52.2.37.

Ingold, T. 1999. Comment on "Animism" Revisited. *Current Anthropology* 40 (Supplement): 81–82.

Ingold, T. 2000. *The Perception of the Environment: Essays on Livelihood, Dwelling and Skill*. London: Routledge.
Ingold, T. 2015. *The Life of Lines*. Abingdon: Routledge.
Iossifidis, M., and L. Garforth. 2022. Reimagining Climate Futures: Reading *Annihilation*. *Geoforum* 137: 248–257. https://doi.org/10.1016/j.geoforum.2021.12.001.
IUGS. 2024. The Anthropocene. International Union of Geological Sciences, March 20. https://www.iugs.org/_files/ugd/f1fc07_40d1a7ed58de458c9f8f24de5e739663.pdf.
Johnson, M.H. 2012. Phenomenological Approaches in Landscape Archaeology. *Annual Review of Anthropology* 41 (1): 269–284.
Joshi, S.T. 2011. Introduction. In *An Epicure in the Terrible: A Centennial Anthology of Essays in Honor of H. P. Lovecraft*, ed. David E. Schultz and S.T. Joshi. New York: Hippocampus Press.
Kruse, F. 2013. *Frozen Assets: British Mining, Exploration, and Geopolitics on Spitsbergen, 1904–53*. Circumpolar Studies 9. Groningen: Arctic Centre, University of Groningen.
Lane, P.J. 2015. Archaeology in the Age of the Anthropocene: A Critical Assessment of Its Scope and Societal Contributions. *Journal of Field Archaeology* 40 (5): 485–498. https://doi.org/10.1179/2042458215Y.0000000022.
Lantto, M. 2020. San Pedro River Archive. *Cultural Geographies* 27 (1): 143–155.
Latour, B. 2018. *Down to Earth: Politics in the New Climatic Regime*. Cambridge: Polity Press.
Lockhurst, R. 2017. The Weird: A Dis/Orientation. *Textual Practice* 31 (6): 1041–1061.
Marr, N., K. Lantto, M. Larsen, K. Judith, S. Brice, J. Phoenix, C. Oliver, O. Mason, and S. Thomas. 2022. Sharing the Field: Reflections of More-Than-Human Field/Work Encounters. *GeoHumanities* 8 (2): 555–585.
Martín Porras, B. 2022. Monumentalising the Monstrous: The Attraction and Repulsion of Monsters in Archaic Greek Art. PhD thesis, King's College London.
Matila, T. 2020. The Ghosts in the Archive: World War Two Photography and Landscapes Crafted by the Nazis in Finland. *Time and Mind* 13 (4): 351–371.
Mittman, A.S. 2012. Introduction: The Impact of Monsters and Monster Studies. In *The Ashgate Research Companion to Monsters and the Monstrous*, ed. A.S. Mittman and P.J. Dendle, 1–14. Farnham: Ashgate.
Morton, T. 2016. *Dark Ecology: For a Logic of Future Coexistence*. New York: Columbia University Press.
Moshenska, G. 2006. The Archaeological Uncanny. *Public Archaeology* 5 (2): 91–99. https://doi.org/10.1179/pua.2006.5.2.91.

Moshenska, G. 2012. M.R. James and the Archaeological Uncanny. *Antiquity* 86 (334): 1192–1201. https://doi.org/10.1017/S0003598X00048341.
Musharbash, Y. 2014. Introduction: Monsters, Anthropology, and Monster Studies. In *Monster Anthropology in Australasia and Beyond*, ed. Y. Musharbash and G.H. Presterudstuen, 1–24. New York: Palgrave Macmillan.
Nordin, J.M. 2020. *The Scandinavian Early Modern World: A Global Historical Archaeology*. Abingdon: Routledge.
Nylander, E.-K. 2023. *From Repatriation to* Rematriation: Dismantling the Attitudes and Potentials Behind the Repatriation of Sámi Heritage. PhD thesis, University of Oulu, Oulu.
Olsen, B., and Þ. Pétursdóttir. 2014. *Ruin Memories: Materiality, Aesthetics and the Archaeology of the Recent Past*. Abingdon: Routledge.
Olsen, B., and Þ. Pétursdóttir. 2021. Writing Things After Discourse. In *After Discourse: Things, Archaeology and Heritage in the 21st Century*, ed. B. Olsen, M. Burström, C. DeSilvey, and Þ. Péturdsóttir, 23–41. Abingdon: Routledge.
Paphitis, T. 2020. Haunted Landscapes: Place, Past and Presence. *Time and Mind* 13 (4): 341–349.
Paphitis, T. 2021. Fielding the Mind in the High North. *Time and Mind* 14 (3): 343–344.
Paphitis, T., R. Norum, and V.-P. Herva, eds. 2021. *Time and Mind* 14 (3): Special Issue on 'Minding Arctic Fields'.
Pétursdóttir, Þ. 2017. Climate Change? Archaeology and Anthropocene. *Archaeological Dialogues* 24 (2): 175–205.
Rae, A. 2020. The Uncanny, the Weird and the Eerie. Hyperobjects and Anthropocenic Modalities in China Miéville's Three Moments of an Explosion. In *Fiction and the Sixth Mass Extinction: Narrative in an Era of Loss*, ed. J. Elmore, 109–132. London: Lexington Books.
Seamon, D. 2018. Merleau-Ponty, Lived Body and Place: Toward a Phenomenology of Human Situatedness. In *Situatedness and Place*, ed. T. Hünefeldt and A. Schlitte, 41–66. Cham: Springer.
Seitsonen, O. 2020. Ruttoa ylätunturissa? Saamelaisten Poronhoitokohteiden Ajoittuminen Enontekiön Yliperällä. *SKAS* 4 (2020): 2–20.
Seitsonen, O. 2021. *Archaeologies of Hitler's Arctic War. Heritage of the Second World War German Military Presence in Finnish Lapland*. Abingdon: Routledge.
Seitsonen, O., and S. Viljanmaa. 2021. Landscapes of Sámi Reindeer Domestication and Pastoralism in the Gilbbesjávri Region, Sápmi, Northernmost Europe ca. 700–1800 A.D. *Journal of Field Archaeology* 46 (3): 172–191.
Seitsonen, O., L.G. Broderick, I. Banks, M. Olafson Lundemo, S. Seitsonen, and V.-P. Herva. 2021. Military Supply, Everyday Demand, and Reindeer: Zooarchaeology of Nazi German Second World War Military Presence in Finnish

Lapland, Northernmost Europe. *International Journal of Osteoarchaeology* 32 (3): 3–17.

Stichelbaut, B., S. Thomas, O. Seitsonen, W. Gheyle, G. De Mulder, V. Hemminki, and G. Plets. (2021). Operation Northern Light: A Remote Sensing Approach to Second World War Conflict Archaeology in Northern Finland (Kilpisjärvi, Enontekiö). In *Conflict Landscapes: Materiality and Meaning in Contested Places*, ed. N. Saunders and P. Cornish, 202–220. Abingdon: Routledge.

Tabas, Brad. 2015. Dark Places: Ecology, Place, and the Metaphysics of Horror Fiction. *Miranda* 11 [online]: 1–17. https://doi.org/10.4000/miranda.7012.

Thomas, S., V.-P. Herva, O. Seitsonen, and E. Koskinen-Koivisto. 2019. Dark Heritage. In C. Smith (Ed.), *Encyclopedia of Global Archaeology*. New York: Springer.

Tilley, C. 1999. *Metaphor and Material Culture*. Oxford: Blackwell.

Tilley, C. 2008. Phenomenological Approaches to Landscape Archaeology. In *Handbook of Landscape Archaeology*, ed. B. David and J. Thomas. London: Routledge.

Tsing, A., H. Swanson, E. Gan, and N. Bubandt. 2017. *Arts of Living on a Damaged Planet: Ghosts and Monsters of the Anthropocene*. Minneapolis: University of Minnesota Press. http://www.jstor.org/stable/10.5749/j.ctt1qft070.

Tuominen, S., and M. Marila. 2021. I Shed Tears, Left, and Forgot: The Common Frog, Mosquitoes, and Grandmother Pine Stayed. In *Heritage Ecologies*, ed. R. Bangstad and D. Pétursdóttir, 369–380. Abingdon: Routledge.

Turnbull, J., B. Platt, and A. Searle. 2022. For a New Weird Geography. *Progress in Human Geography* 46 (5): 1207–1231.

Ulstein, G. 2021. Weird Fiction in a Warming World: A Reading Strategy for the Anthropocene. PhD thesis, University of Gent.

Ureta, S., and P. Flores. 2018. Don't Wake up the Dragon! Monstrous Geontologies in a Mining Waste Impoundment. *Environment and Planning d: Society and Space* 36 (6): 1063–1080. https://doi.org/10.1177/0263775818780373.

Varnajot, A., and E. Salim. 2024. The Hauntology of Climate Change: Glacier Retreat and Dark Tourism. *Tourism Geographies*: 1–18 (online first). https://doi.org/10.1080/14616688.2024.2328607.

Vermeulen, P. 2020. *Literature and the Anthropocene*. Abingdon: Routledge.

Viveiros de Castro, E. 1998. Cosmological Deixis and Amerindian Perspectivism. *Journal of the Royal Anthropological Institute* 4 (3): 469–488.

Wallenhorst, N. 2023. *A Critical Theory for the Anthropocene*. Cham: Springer. https://doi.org/10.1007/978-3-031-37738-9.

Wengrow, D. 2014. *The Origins of Monsters: Image and Cognition in the First Age of Mechanical Reproduction*. Princeton: Princeton University Press.
Wright, S. 2012. Emotional Geographies of Development. *Third World Quarterly* 33 (6): 1113–1127. https://doi.org/10.1080/01436597.2012.681500.

Open Access This chapter is licensed under the terms of the Creative Commons Attribution 4.0 International License (http://creativecommons.org/licenses/by/4.0/), which permits use, sharing, adaptation, distribution and reproduction in any medium or format, as long as you give appropriate credit to the original author(s) and the source, provide a link to the Creative Commons license and indicate if changes were made.

The images or other third party material in this chapter are included in the chapter's Creative Commons license, unless indicated otherwise in a credit line to the material. If material is not included in the chapter's Creative Commons license and your intended use is not permitted by statutory regulation or exceeds the permitted use, you will need to obtain permission directly from the copyright holder.

CHAPTER 3

The "Glacier"

We were all excited about this fieldwork opportunity well beforehand. Even though we considered ourselves to be reasonably familiar with Lapland, Ritničohkka with its "glacier" seemed to bring something new to the land, which, generally speaking, never ceases to surprise. We had a rather clear idea of our actual objectives: mainly, the collection and documentation of recently thawed reindeer bones, which could open up an osteological library going back centuries. This goal to collect reindeer bone samples was linked to the broader aim of analysing the reindeer diet and mobility in order to make sense of both human and non-human landscape use over time (Fjellström et al. 2022; Seitsonen and Fjellström 2022; Salmi and Seitsonen 2022). But at the same time, we couldn't help sensing that a more unexpected adventure awaited us here as well (Fig. 3.1).

The five of us have done fieldwork and lived in Lapland or elsewhere in the European High North for various periods and, between us, had been studying Arctic pasts and presents for some two decades. Our prior research approached the landscapes and inhabitants of Lapland

---

The original version of the chapter has been revised. Several passages amended to include proper attribution. A correction to this chapter can be found at
https://doi.org/10.1007/978-3-031-85016-5_10

© The Author(s) 2025, corrected publication 2025
V. Herva et al., *Weirding Landscapes*, Arctic Encounters,
https://doi.org/10.1007/978-3-031-85016-5_3

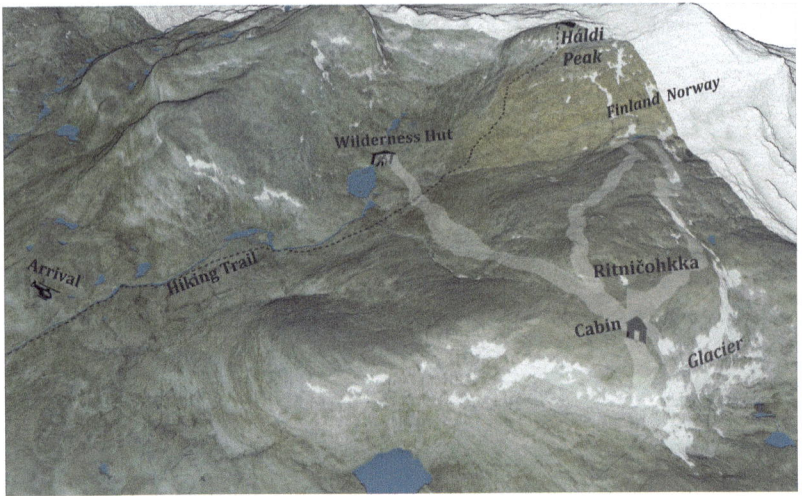

Fig. 3.1 A bird's eye view of our expedition's journey from the south-east. The extent of the snowfield in the aerial imagery is as it was in 2012 (National Land Survey of Finland, orthophoto dated 2012 and 2 m resolution digital terrain model [DTM] and Norwegian Mapping Authority, 1 m DTM). The light grey shading shows our route as recorded by GPS navigators

from myriad angles and using a range of methods, from bioarchaeology and GIS analysis to the phenomenology of landscape and sensory ethnography. We have worked extensively with ecological, biological, archaeological, historical, archival, anthropological and folklore data. Our research has encompassed long-term landscape formation processes and temporalities, Sámi pastoralist landscapes, spiritual engagements with northern landscapes through time, and how Lapland has been perceived and represented by outsiders, to mention but a few themes and topics (e.g. Herva et al. 2020; Herva and Ylimaunu 2009; Herva and Seitsonen 2020; Hakonen 2021; Fjellström et al. 2022; Seitsonen and Fjellström 2022).

Yet despite having conducted fieldwork in and around the Kilpisjärvi region for some time, none of us had been to the particular area of Ritničohkka. The unfamiliar, remote nature of Ritničohkka rendered it exciting, and admittedly exotic, to each of us—something further amplified by its renown as the site of the only putative glacier-like formation in Finland, next to the country's highest peak.

The basic idea for our fieldwork was straightforward enough: to survey this snowfield, which we presumed to be in the process of melting, and to snoop around to see what archaeologically interesting material it may have released. Various forms of "glacier archaeology" have been carried out in recent years in neighbouring Norway and Sweden, as well as elsewhere in the world. Hailing from the Alps that run along the Austrian-Italian border, "Ötzi" is perhaps the most famous example of what melting glaciers can bring to the surface.

As we did not expect to find a Finnish Ötzi, our ambitions were somewhat more modest: reindeer bones. These would be useful for analysis in further research. And yet, as one never knows what one might find when a glacier melts into thin air, there was a frisson of excitement and anticipation in the days leading up to our departure. Indeed, even we, as relative insiders, all living in Finland or Sweden, each held particular expectations of the place where we were heading, even if they were not completely aligned with the typical, long-standing, external perceptions of the High North as a strange and magical world (Herva et al. 2020). Our aerial arrival at the rock outcrop a top Ritničohkka only heightened these expectations and perceptions.

3.1 A Flying Transition

"In many ways", as Graves-Brown and Schofield have written regarding the forms of mobility involved in archaeological surveying, "flying is the ultimate example of the way in which a change in modality can radically alter perspective" (Graves-Brown and Schofield 2020: 7). Our northern journey began with an overnight train, followed by a several-hour car ride from southern Finland to the village of Kilpisjärvi (Gilbbesjávri in Northern Sámi), where we took a few days to gear up and prepare for our trip further north. We settled into rooms at the University of Helsinki owned Biological Research Station, the main port of call for researchers across disciplines who base themselves here for a wide range of scientific excavations, explorations, excursions and incursions in the area.

At the station, we met with some colleagues and gave a short lecture to a small audience of scientists and locals about our impending journey and what we were hoping to find. We spent some time turning on our email auto-replies, checking the field equipment once more, and poring over the Ordnance Survey maps of the various trails and routes we were to walk. We double-checked the agreed-upon helicopter schedule at the local helipad and purchased necessary last-minute provisions and equipment for the planned five days out in the fjells. This equipment included gear for all weather; food, mainly add-water soup packets and plastic jugs of potable water, packets of elk jerky, freeze-dried desserts and even a box of wine (anticipating a celebration if something interesting was found); our compasses and maps, supplemented by GPS navigators and mobile phones, as well as a satellite phone and GPS rescue beacons, just in case; sealable find bags, pens, notebooks; a drone, a 360-degree camera and miniaturized 3D scanning equipment. We found ourselves with a cumbersome load, which only continued to expand as we put more and more trust in the helicopter's ability to carry it.

Once we were sufficiently packed and prepared for what lay ahead, we drove to the Kilpisjärvi Heliport. There, from June to early September, the company Heliflite provides transport to groups of hikers, hunters and fly-fishers, while also carrying out the odd search-and-rescue operation across Northern Finland and Sweden. We had been using their services during our earlier fieldwork and were already familiar with the owner and the pilot. As we drove in, we were greeted by the distant whirring, then buzzing and finally whoosh-whooshing of an incoming turbine engine.

The company operates a fleet of four helicopters; the one in operation that day at Kilpisjärvi was a Eurocopter AS 350B2, a turbine chopper commonly used for charter flights and aerial work in rugged conditions, with room for six and a cruise speed of around 225 km/h. We loaded our 15 bags of equipment, food and supplies onto the grated steel carrying baskets lining the helicopter's fuselage on both sides, piled in, buckled ourselves in and put on the intercom headsets. The helicopter took off swiftly, putting increasing distance between us and the shrinking human paraphernalia of houses and car lots. We lost ourselves in the transition from the familiar to the unaccustomed.

The 30-minute helicopter flight from Kilpisjärvi to Ritničohkka was deeply defamiliarizing. Yet, it served as an orienting experience for our perceptions and receptions of the landscapes of Ritničohkka in the days ahead. While we knew, both intellectually and from first-hand experience, that this was a land comprised primarily of very rocky tundra that is difficult to traverse by foot, catching a glimpse of the vast, undulating and largely monotonous Alpine tundra landscape from the vantage point of

150 metres above the ground proved deceitful, as large stones and boulders seem no larger than pebbles. Also, a sense of disorientation—the first of many on our journey—of seeing an otherwise familiar landscape from a wholly new angle, was further deepened once we got to exploring Ritničohkka on foot. The helicopter flight itself provided a re-scaling of our orientation, as it revealed the vastness of the landscape, dominated by an unending boulder field, where for the most part only reindeer and their herders roam.

The perspective granted us by the helicopter was significant because it facilitated viewing the landscape with a perspectival, bird's eye view but from a height above ground that was still low enough for us to be able to discern the forms and features of the barren land. In the August sunlight, this gave the undulating hills something of a dreamlike quality, as they extended seemingly to infinity. This vast boulder field we saw makes up much of the upper north-western "arm" of Finland and extends across both sides of the border into Norway and Sweden. Even in geological reports this area is sometimes referred to as a "moonscape" (Lehtovaara 1995: 58), with thin layers of lichen and moss typically being the only visible signs of life. Unlike passenger jets which cruise at higher altitudes, or even general aviation planes which are less easy to manoeuvre and typically maintain higher flight paths, the lower-flying helicopters, skidding along and following the undulations of the terrain at a low level, generate peculiar sensations of both being in, and not being in, a given landscape (Fig. 3.2). This has resonances with shamanistic "flying", common to northern cosmologies, which also speaks to the process of perceiving a given reality in different ways and from different angles (e.g. Herva and Lahelma 2020). The various, variegated qualities of the landscape we saw and sensed from the helicopter came to play a marked role in how we perceived and related to the land—particularly after we landed on it.

As we held our handheld GPS and mobile phones, following the map readouts on them, Ritničohkka slowly came into view. The helicopter began preparing for its descent. The peak was identifiable not by any particular outcrop of rock or land but by a radio tower and two small cabin structures. Our bird's eye view onto the fjell readily emphasized the isolated location of the cabins. The nearest settlements, Kilpisjärvi among them, are located 40 kilometres away as the crow flies, a distance that looks very reachable on the maps we had. But traversing the hills actually would have taken several days across rather difficult terrain. With this in mind, gazing at the surrounding land from the air underlined our

Fig. 3.2 Soaring over the barren reindeer country

sense of isolation and distance from our accustomed world, even though we were, in fact, just a few kilometres away from a popular summertime hiking trail from Kilpisjärvi to Háldi, the fjell adjacent to Ritničohkka and the absolute highest elevation in Finland.

The pilot began assessing for a potential spot to land the machine. On our first approach at the outcrop of exposed rocks, the copter righted itself, tilting back in controlled fashion and slowing down, slightly overshooting the mountaintop as the pilot guided us into a full turn followed by a swiftly descending hover. Holding us just above the ground, he eased the front of the right-hand side skid onto a level rock. The rest of the craft continued to hover, though only just, its right side clinging to the rock like a boat to an anchor. The helicopter finally landed fully on the jagged boulders with a hefty, solid thud. The pilot's smirk suggested he had done this plenty of times. Outside the cockpit, up a gently sloping gradient, the cabins and their adjacent abandoned radio antenna tower stood unimpressed. The sky above was clear and the view extended over the surrounding mountains all the way to the far-off horizon towards the Arctic Ocean.

As the chopper lifted and drifted off into the horizon, our sense of isolation magnified.

3.2 The Fjells of the North: Political Geographies, Mindscapes and Mythical Resonances

This wilderness region in Finland's north-western arm is called "the Upper End". It is a *vidda*, a vast expanse empty of roads and permanent settlements, rife with folk and popular mythology and mythical resonances. The landscape here consists of three primary natural formations: steep-sided rocky fjells, mountain plateaus (covered by little or no vegetation or the odd spot of mountain birch) and river valleys and lakeshores covered with brush. Grey lichen is the region's foremost living entity, as the weather in the uppermost reaches is hostile even to moss (Fig. 3.3). The area is typically encased in clouds for most of the year, with summertime providing some respite. Illustratively, it is very hard to get a good series of satellite images from this region as it is so often overcast. Snow may fall at any time of the year here and often lingers year round.

The border between Finland, Norway and Sweden, which runs along the watershed and the highest peaks, was established along these ridges in 1734 and has since then been crossed relatively freely by flora and fauna, including humans. On the Norwegian side, melting ice and snow empties into the Norwegian Sea, while across the border it flows down to the Tornio-Muonionjoki river system and finally into the Gulf of Bothnia in the Baltic Sea. Human settlement, when not nomadic, has always been spartan here, although during years of the great famine in Finland (1866–1868), Finns migrated in sizable numbers across the northernmost stretches of the country to the north Norwegian coast, resulting in the growth of a Finnish minority there, known as the Kvens.

The main mountain of importance here is Háldi (Fi. Haltitunturi) (Fig. 3.3). A spur of the Ráisduattarháldi fjell that rises 1326 metres above sea level, its highest elevation on the Finnish side is 1324 metres, which holds the accolade of being Finland's highest elevation. The altitude is admittedly rather diminutive when compared to other mountains in Europe, rising only a little more than 400 metres from the surrounding landscape, but for Finland, a low-lying country with no proper mountains to speak of, the fjell is a source of pride and represents the only fragment

Fig. 3.3 View of the Háldi summit, the highest point in Finland

of the majestic Scandinavian mountain range in the country. Never mind that the actual highest summit of this Háldi range is, in fact, located about a kilometre across the border on the Norwegian side, where the Ráisduattarháldi soars to 1361 metres. Even the actual peak of Háldi is a few dozen metres across the border, making Finland's highest elevation a slope just within reach of the summit. Thus, the adjacent Ritničohkka, at 1317 metres, is actually the highest peak in Finland. Nevertheless, it is virtually unknown to most Finns.

To state officials, the land here is known by a recently designated moniker, the Háldi Transboundary Area. This cooperative management area formed in 2020 consists of the Käsivarsi Wilderness Area (in Finland) and the adjacent Reisa National Park and Ráisduottarháldi Protected Landscape (in Norway). In order to climb this peak, most hikers follow a 55 kilometre path due north-east from Kilpisjärvi for three or four days. While there is a much shorter access route from the Norwegian side, passing through Reisa National Park, most Finns take the Kilpisjärvi route. Indeed, it is something of a tradition among wilderness-oriented trekking Finns to hike up to Háldičohkka. Nearby Kilpisjärvi is the famous border marker, the Three Nations' Border Point, where the official state territories of Finland, Norway and Sweden meet at a yellow-painted round cement block situated in the middle of a small lake.

For the Sámi, both Ritničohkka and Háldi hold meanings that are intimately tied to the northern relational and reciprocal ways of seeing, using, feeling and knowing the land. Traditional Sámi worldviews emphasize the deep corporeal and interrelated unity of both non-human and human actors, including, for instance, various landmarks, landscapes, spirits and animals (e.g. Ruotsala 2002; Näkkäläjärvi 2013; Magga and Tervaniemi 2018). In these relations, all breathing and non-breathing, inanimate and animate, invisible and visible entities have their own place as parts of a holistic, comprehensive and dialogic "sentient ecology" (Anderson 2000: 116–117). Such a world derives its meanings from the "*unfolding encounters* with other more or less powerful actors" (Joks et al. 2020: 308, original emphasis), and land itself is seen and approached by traditional Sámi as an actor in a reciprocal field of interlinked connections. It unfolds and presents itself and the diverse meanings attached to it as humans traverse in and as part of it along their habitual and ancestral meshwork of connections, their "lines of becoming" (Hillier 2017; Mazzullo and Ingold 2008).

This intimate connection to land is well illustrated by the example of Sámi responses to the destruction of their homes and infrastructure by German troops across the North during the Lapland War towards the end of the Second World War (1944–1945). What mattered to most Sámi was not whether their houses remained intact but that they still had access to their "own land". Lost buildings and other property could be replaced and lives reconvened so long as their land was there (Seitsonen and Koskinen-Koivisto 2017). This same logic applied particularly painfully when the vast reservoir lakes of Lokka and Porttipahta were built in Soađegilli (Fi. Sodankylä), inundating Sámi ancestral pasturelands and villages and removing locals from their lands forever. Severed from their own past and heritage, many Sámi experienced severe feelings of rootlessness and vulnerability as a result of their forced mobility (e.g. Kauhanen 2024: 341). Even though the displaced Sámi were resettled in the vicinity by the state, the loss of land and the metamorphosis of the landscape resulted in profound transgenerational effects, leading to marginalization, alcoholism and other severe social problems (Kauhanen

2024). This process of losing the ancestral lands is memorialized by the local Sámi as the "re-destruction of Sompio", the first destruction referring to the burning of the villages by the Germans towards the end of the Second World War in 1944 and the re-destruction referring to the Finnish State's colonial reservoir endeavours in the 1960s and 1970s.

Ritničohkka has loomed large as an integral component of the *boazomeahcci* (North Sámi, NS; literally "reindeer forest") pasturelands of Sámi herders and their (semi-)domesticated reindeer herds for centuries— and even longer for the wild reindeer inhabiting the area for millennia (Seitsonen and Viljanmaa 2021). Perennial patches of snow and ice, *jassa* in the North Sámi language, are an important part of both the reindeer's and the reindeer herders' seasonal land use (e.g. Pilø et al. 2018; Taylor et al. 2019; Seitsonen et al. forthcoming). The meltwaters that run down from the permanent fields of ice and snow have long enriched the sparse tundra vegetation of the rocky slopes below them, offering ample lichen pasture for both wild and domestic reindeer. The glaciers and permanent snow patches also offer animals much needed respite from the plaguing swarms of mosquitoes that disturb groups of reindeer and humans alike during the summer months in the North. This has rendered the snowfield a good hunting ground for past hunters of wild reindeer, as well as part of an important seasonal pasture area for the animals themselves.

According to local reindeer herders, Háldi is known in the local vernacular as the "mother of reindeer". This underlines its importance as a springtime reindeer birthing area. In earlier times, it was situated along the herders' seasonal mobility route from their winter to summer pastures. After cross-border migration of the Sámi was restricted by nation states in the later part of the 1800s, the region was redesignated as a summertime pasture of a local *siida* (a traditional Sámi herding cooperative based on kinship, territory and cooperation) (Seitsonen and Viljanmaa 2021: 186).

Háldi is featured in a beautiful joik, a traditional form of Sámi song, which the head of the local reindeer herders, Juha Tornensis, performed for us one evening in his backyard hut in Kilpisjärvi during a prior visit. As he described it, a joik performer does not create the song himself but instead mediates it from his surrounding environment. However, as Juha continued to explain, this specific joik of Háldi must always be sung judiciously and not too frequently, as it usually brings bad weather along with it. Indeed, we witnessed the aftermath of this premonition ourselves, when the day after Juha's performance our field surveys were hindered by a massive rainstorm pelting down from the North.

The Háldi-Ritničohkka area is frequented by herders and their families throughout the year, including during the late summer calf earmarking. Indeed, on the helicopter ride to Ritničohkka, we flew past a busy seasonal Sámi herding camp of the local *siida*. The camps are demarcated by the characteristic scattering of the families' lightweight *lávvus* (teepee-like tents), their ATVs parked alongside them, and by the nearby meandering reindeer fences that direct the movements of their animals. These types of seasonal settlements are important nodes, *báikit* (NS; plural, "places"), that regulate the annual movements of herders following the reindeer in their various *meahccit* (NS; plural, "forests"; Joks et al. 2020; Länsman 2004: 99; Schanche 2002).

Archaeologically and biologically, past Sámi sites are typified by vegetational changes, often with meadow-like flush foliage that distinguishes them from the surrounding areas, as their soils are enriched by recurring long-term use by animals and humans (e.g. Itkonen 1948: 274; Karlsson 2006: 163; Seitsonen and Égüez 2022). What is notable in this context is the fluidity with which the *báikit* and *meahccit* are defined, as their boundaries shift and overlap seasonally and temporally, yet they are all part of the all-inclusive home landscape (Länsman 2004: 99). This also clearly illustrates how any strict nature/culture dichotomies in Western and modern societies are meaningless in the holistic Sámi worldview; as one of our informants aptly commented to us: "Home is all our fjells".

3.3 The Discovery of the Glacier

Carrying our backpacks and duffle bags of equipment from the helicopter landing spot up to the cabin gave us our first and disorienting sense of the rocky ground that would occupy our imagination over the coming days. Following a protracted episode of entering the cabin (see Sect. 5.1), we immediately set off to explore Ritničohkka's eastern slope, as we wanted to take full advantage of the good weather, which might change in an instant at those elevations. This was where the main bulk of the snowfield was located. It was a sunny and surprisingly warm August afternoon, and though moving over the rocky slope was tedious, Ritničohkka felt friendly and welcoming. At the same time, several of us were wondering how we would in fact come to understand this place and its many facets.

Slowly we descended along the slope, in an open-ended landscape that dwarfed us with its grandeur. The same boulder field extended as far as the eye could see and beyond. The foothold consisted of loose rocks and boulders, on terrain where a level surface is a rare luxury. It took a while, despite our prior experience, to get newly accustomed to the unique topography and visual signals of the discomposed rocks. We could definitely make out signs of a recent melt: the rocks farther away were of a different shade than those at our feet. The colour difference was caused by a thin veil of creeping lichen. It became obvious that the bare rock must mark the location of recently melted parts of the snowfield. The first discoveries of reindeer bone from the hollows in the boulder field suggested as much.

The steep slope which we were now traversing horizontally towards the north was cut by slightly elevated vertical ridgelines, which blocked the view of the rest of the eastern slope. There, beyond our vision, is where the snowfield must have receded. Ascending the crest would offer us a clearer view. Choosing our steps carefully, to avoid injuries on our first day, one by one we crossed the ridge and witnessed our target.

The Ritničohkka "glacier" was reported in 1892 in a geological report by Hugo Stjernvall. Seen in aerial photographs taken between 1960 and 1991, it measured on average some 5 kilometres long and about 600 metres wide. This formation was over 6.5 metres thick when Finnish geologists drilled an ice core from it in 1993. A report in 2000 concluded that the snowfield was "stable under the current climatic regime" (Hirvas et al. 2000), but already during their next visit in 2005 researchers expressed alarm at its rapid shrinking (Hirvas et al. 2005).

Ritničohkka's snowfield made the news in 2021 after huge chunks of it tumbled down the slope in an avalanche, an event witnessed and filmed by a Finnish off-piste snowboarding crew. The history of the ice field before it appeared in archival texts is unclear, as earlier research suggested that it might not be much older than its earliest reference by Stjernvall, dating to the Little Ice Age (c. 1300–1850 A.D.). Our research would go on to prove that the age of the snowfield, in some form or another, was at least two thousand years, based on our radiocarbon-dated archaeological finds, such as reindeer bones (Seitsonen et al. forthcoming).

This was technically not the first time we saw our target area, as the helicopter had earlier overshot the mountaintop and made a slow turn over where we now stood. But now we all saw clearly what little there was to see: pitifully diminished patches of snow clinging to the boulder-strewn slope. Not a glacier or a snowfield as we knew it, no. We even told each other so on the intercom, asking whether anyone saw any ice, but dismissing what we saw with our own eyes, thinking that perhaps the snowfield was hiding somewhere beyond our field of vision. There had to be, we told ourselves, more snow and ice somewhere, behind some crest that had prevented us from seeing it in the helicopter.

There wasn't. Only scattered patches of snow here and there, meltwater dripping and trickling, tiny runnels merging into audible streams flowing below the rocks. At their core, the largest left-over patches of snow and ice measured a few dozen metres across and at most 2 metres thick, but they were receding and losing mass at an alarming rate. Bones of dead animals, mostly apparent remains of reindeer (as the post-fieldwork analyses confirmed), were unceremoniously collected and documented by our team from various places of recent melt (Figs. 3.4, 3.5 and 3.6).

Each of us undoubtedly had our own varying ideas of what the snowfield or a glacier might look like. How thick or vast it might be, what the landscape around it would look like, what we would find on or in it. Our imaginations and anticipations of the snowfield's topography were formed and framed by our experience of the environmental change taking place in Finland and Lapland over the past several years, and by the place of glaciers in the social imaginary more broadly.

Fig. 3.4 With each uncertain step, the team scavenged the remains of the vaunted glacier

Fig. 3.5 Another glacial refuge not long for this world. Note the difference between the recently uncovered vegetation-free light grey rocks and the darker shades in the distance

Fig. 3.6 Markus exploring the last subglacial

We had heard, apocryphally, of the avalanche several years prior that was said to have obliterated what remained of the snowfield, but this story was based on hearsay from locals and tourism entrepreneurs, and its veracity had remained unclear. In any event, the remoteness of the spaces around Ritničohkka (and their uninhabited, or at least unvisited, nature) were such that all of us were convinced that, even if others in the past had been unable to find traces of the snowfield or had even seen it disappearing, surely we would still be able to locate its receding core. We were not.

This was *not* a glacier or a snowfield. It was certainly far from a continuous sheet of ice, far from how had imagined it in our mind's eye, based on the photographs from the 1990s and early 2000s we had viewed. There wasn't any single, uniform and connected stretch of snow and ice but only patchy spots spread across various folds of the landscape, which mostly consisted of loose stones of various sizes atop the bedrock. The flats of snow looked out of place, dwarfed as they were by the vast rocky plateaus and undulating fjells that stretched far and wide, well out of our peripheral vision. It was almost as though the snow and ice had been dropped arbitrarily from high above onto various stony outcrops. It did not seem as though they were autochthonous parts, bound to the landscape and its histories. The presence of ice and snow felt decidedly alien.

Walking on any live glacier can be a challenging experience: the surface is often uneven and slippery, and the deceitful snow-cover concealing hidden crevasses contrasts with the solid ground. But there was rather little of this to fear when we walked up onto the perimeter of Ritničohkka's ice. The step up was too diminutive, the edges flattened too visibly, and we knew that vast crevices, sinkholes or hidden canyons were unlikely in the underlying boulder field. All this gave us confidence that we could walk about on the glacier's remains without risk of being harmed by anything unseen, unpredictable or unknowable. The previous time that scientists visited this place in the early 2000s, the situation was vastly different, with over 6.5-metre-deep deposits of layered ice and permafrost down to a depth of 15 metres, as shown by electrical conductivity studies and the ice core drilled by geologists (Hirvas et al. 2000; Vanhala et al. 2005).

The patches of ice and snow were mostly curiously pockmarked with handbag-sized circular divots, looking as though they had been trampled on by someone just a few hours prior. In fact, the divots in the surface ice had likely formed gradually because of the melt or winds or both. We also encountered several patches of pure white snow, symmetric in some parts, asymmetric in others. In short order, we were crouching and kneeling on their margins to grasp the general feel of the last bits of disappearing glacier.

As we did so, the snow patches began to feel sentient, as if we were keeping watch over the remains of a dead body, with the eerie motionlessness of the last bits and pieces of a corpse that was not moving but somehow felt as though it was about to. The faint background dribble of meltwater provided the soundtrack of a dying planet. This snow and ice was no longer part of a living snowfield or glacier. Neither were the patches; they were dead bodies, definitely not "breathing, moving ever slowly across the valley like an old, waking body", as Anna Ellegaard Buhl (2023) describes a proper *dying* glacier, as opposed to a dead one. These were left-over body parts, the discarded remnants of a once-frozen being, picked apart by the vultures of time and heat.

3.4 What Is a Glacier?

Glaciers, it seems, are having a moment. One Swiss study suggests that by 2050, about half of the planet's glacier volume will disappear. In the Alps, it is estimated that more than 90% of the region's glacier volume could be lost by 2100. The UN General Assembly declared 2025 the International Year of Glacier Preservation; 21 March has been designated the World Day of Glaciers. Perhaps the importance of glaciers need not be overstated, but let us indulge for a moment. The flow of glaciers has sculpted mountains and carved valleys for most of the Earth's history (and continues to do so), giving form to much of what we see today on our planet. Glacial ice is the largest reservoir of freshwater on Earth, sustaining stream habitats for plants, animals and humans. Glacier melt delivers nutrients into lakes, rivers and oceans, forming phytoplankton and, in turn, entire marine food chains.

But what is a glacier, ontologically speaking? In the concrete, linguistic sense of the term, a glacier is a slowly moving mass or river of ice formed by the accumulation and compaction of snow on mountains or near the poles. These persistent bodies of dense ice, constantly moving under their own weight, form in places where the accumulation of snow exceeds its ablation over many years, often centuries. But what does a glacier do? What does a glacier mean? And how might we work with, or think through, glaciers?

As an integral component of the hydrological cycle, glaciers are of fundamental importance to climate, biodiversity and the environment. Glaciers are, to be sure, geophysical indicators and bellwethers of planetary climate effects across the Earth. They act, for instance, as fragile and friable thermometers for the planet's fluctuating temperature. Glaciers have thus also become poster children of global environmental change and of human action for or against such change. While they may not sit in quite the same boat as the polar bear, the seal, the whale or other charismatic megafauna whose cute, pleading images galvanize action around the environment, glaciers loom large in the social imaginary of environmental and climate change.

Glaciers have in recent decades been shown to hold more than their fair share of agency (see e.g. Cruikshank 2005; Orlove et al. 2008; Carey 2007; Jackson 2015; Sörlin 2015). Indeed, despite positivist arguments to the contrary, ice does much more than just melt (Pollack 2009: 114). The normative framework through which Western societies have conceptualized and dealt with ice (e.g. through the science of empirical glaciology) is "not a neutral representation of nature" (Carey et al. 2016: 787). Building on calls to "unsettle and challenge dominant assumptions" that typically underscore how Euro-American knowledge is produced (Harris 2015, cited in Carey et al. 2016), we suggest that glaciers, their representations and their understandings within the social imaginaries of climate change be picked apart in order to determine what we, as humans, can (or should) do with glaciers.

In a recent essay, Zachary Provant and Mark Carey (2022) have critically interrogated the latter-day phenomenon of glacier funerals, posing the urgent question, "Do glaciers just die, or are they killed?" In 2016, Icelandic Meteorological Office glaciologist Oddur Sigurðsson publicized the death of Okjökull, the first glacier officially recognized to be lost to human hands—cause of death: 'excessive heat' and 'humans'—and the first of Iceland's several hundred glaciers to lose its official glacier status (Provant & Carey 2022). Several years later, a hundred odd people hiked up the volcano where Okjökull once lay to attend a belated memorial, led by anthropologists Cymene Howe and Dominic Boyer (Howe & Boyer 2024). The 2019 hike was accompanied by speeches, a moment of silence and the dedication of a small bronze plaque mounted on a rock (Kenny 2021).

Scholars and activists followed suit with other glaciers around the globe. In Switzerland, the commemoration of Pizol Glacier's passing was held in 2019 alongside an UN climate conference; Oregon's Clark Glacier memorial a year later was streamed on Instagram (Provant & Carey 2022). As Provant and Carey (2022) have discussed, during the service the director of the Clark event, Anders Carlson, having collected some of the deceased glacier's meltwater, "put the vial into a coffin, draped it with black cloth and positioned it in front of the stark-white state capitol building". The authors insightfully note that, while Carlson's words indeed put a spotlight on humans' responsibility for glacial melt, calling for change, the resulting press "focused more on the scene of mourning than the very thing the organizer had intended: actions and solutions" (Provant & Carey 2022).

The rite of passage of the glacier funeral, Provant and Carey further argue, co-opts affective moments of "melancholia, urgency, and spectacle" (2022). Nevertheless, in so doing, this ritual rarely directly reference anthropogenic climate change, nor does it speak to the *anthropocenic* injustice that "the people living close to glaciers are the ones who contribute the least to their deaths but are most at risk from their loss" (Provant & Carey 2022) (Figs. 3.7 and 3.8).

3.5 Weirded Rocks of a Post-glacial World

The more we explored the fjell's surrounds and the area where the snowfield had been, the more the rocky terrain grew imposing and fascinating—as well as uncanny. The ground that had in less than a decade been exposed by the melting ice looked and felt subtly different from the rest of the landscape. This was not solely due to the absence of lichen and the different shades it produced—the recently exposed rocks were also unsettled and alive. With the glacier gone, our attention turned to the rockscape. Yet, a ghostly and haunting presence lingered in the landscape, a contorted figment of a now-fictitious glacier and a transitioning landscape that affected our psyche, with the remains of snow and ice and the oozing of the meltwater like an "ectoplasm" of the glacier's lingering spiritual presence (cf. Kaplan 2003).

The jagged rocks that comprise the visible surface of Ritničohkka are loosely strewn about the land, paving a rather dangerous surface on which one must walk. As any step could shift or roll a stone and thereby twist an ankle, and as sharp rock edges could easily cut through clothes and

3 THE "GLACIER" 63

Fig. 3.7 A glacier that once was: remains of Mer de Glace (Eng. "Sea of Ice") in Chamonix, French Alps. Notice the sign showing the 1990 surface level of the glacier. Even in the early 2000s, the valley was still filled with glacial ice to the upper limits shown by the open soil, but since then it has been shrinking at an ever-increasing speed of some 40 metres per year, losing about 80 metres in thickness. At the bottom of the steep ravine, one can see the pitiful remains of the once massive ice core

skin, much of our attention and focus during the field surveying turned to the stones themselves. Boulder fields are quite common phenomena in Finland, and they are often places of folklore and tales, commonly referred to as "Devil's Fields" (Fi. *pirun pelto*) and considered the agricultural antithesis of fertile soil. In general, the boulder fields at lower elevations consist of more rounded stones and are thus more proximate in time to ancient shorelines, an indication that the fields were at one point below sea level, where currents and coastal waves over time churned and polished the stones. Unlike most of Finland, 76% of which has at some point during the Holocene been under the sea level, 70% of Finnish

Fig. 3.8 The pockmarked face of a melting snow patch, jassa, on the slope of Gieddečohkka (Fi. Etu-Halti), with the summit of Háldi on the far right

Lapland has remained above the sea, which has had a marked effect on the region's rugged landscape.

Ritničohkka consists of two main geological formations. The top and the north-western parts are dominated by igneous (heat-formed) gabbro rock. Gabbro generally forms beneath oceans, when lava cools slowly under great pressure. The steeper eastern slope is characterized by banded metamorphic gneiss (Sipilä 1992), which forms under even greater pressure, such as at the deep collision points of continental sheets, where the stone remelts and its mineral structure is rearranged into uneven layers. Their sheer presence embodying deep time, both types of rock were perceptually most notable for their dull greyness, which provided the main colour scheme of the landscape, contrasted by the blue of the sky.

The rock formations of Ritničohkka have previously been defined as Silurian, roughly 438–422 million years old. The Silurian is a geological period when much of the Earth's landmass was pushed down by the oceans. Ritničohkka would have been just south of the equator, at the northern rim of the ancient continent of Baltica. From there, the region has crawled north-eastward—something it continues to do to this day, at a speed of 2 centimetres per year, according to precise GPS measurements (Lahtinen et al. 2022).

During the Silurian period, Baltica collided near the equator with the larger continent of Laurentia, later to become North America. As the two primaeval continents wrestled with one another, Baltica slipped underneath Laurentia, thrusting onto magmatic rocks tectonic formations of tilted folds known as nappes, which have survived as the Scandes (or Scandinavian Mountains), Scandinavia's largest mountain range (at 1762 kilometres, it is also the longest range in Europe), forming the spine of the Scandinavian peninsula (Selley et al. 2005; Majka et al. 2012). This explains the relatively young Silurian intrusion over the bedrock of Finland, which one catches glimpses of from time to time. This epoch of mountain forming, known as the Caledonian orogeny, represented at the very tip of Finland's north-western arm, is considerably younger than the rest of Finland, where the bedrock is pre-Cambrian (i.e. 1–4 billion years old) and thus predates the entire fossil record of the world (Kohonen and Rämö 2005; Luukas and Kohonen 2021).

The bedrock on the eastern slope has managed to surface partly as hanging gargoylesque figures, with cleaved rocks strewn underneath due to foliation—a cleavage of deformation and metamorphism caused by expanding and contracting pressure and heat over the years. Here, we encountered several boulders of rust-coloured troctolite, their place of origin easily identifiable as the neighbouring rust-coloured Háldi on the other side of Ritničohkka's peak. The cleaving of bedrock in such immense quantities and the carrying of stones over a mountaintop was evocative of the awesome transformative—transportative—power of ice. Twelve thousand years ago, most of Fennoscandia was covered by a massive ice sheet. One of its last refuges, before it finally dissolved, was here, in north-western Lapland. At its thickest, the ice mass that covered the Ritničohkka-Háldi region measured approximately 1 kilometre thick (Patton et al. 2016), a continental glacier of magnificent proportions.

The newly exposed ground on which we stood felt lifeless yet animate. As nothing had thus far been able to grow on these rocks, they looked like they had cleaved off recently and were still trying to find their place. This odd combination of life and lifelessness—tangible rocky presences and an otherworldly sense of latent, intangible "awareness"—was embedded in a landscape that was simultaneously ancient and timeless, and yet constantly changing—something that became more evident later on in our journey. This conflicting sensation was just one of the many dissonances that mediated our perception, our experience and our sense of this remote land, sparking no small amount of rumination on the character and spirit of the place we had arrived at.

We started to notice various intriguing characteristics of the rockscape as we became more immersed in the place; while they first seemed to be characterized by monotonousness, a richness and diversity gradually opened up as we became more absorbed in the landscape. Little by little, we began to spot all kinds of curious features. The deep past of Ritničohkka is visible in the form of the twisted metamorphic gneiss bands along the eastern slope. Blocky outcrops grew out of the gneiss, with layered bands in the bedrock reaching for the sky at 45-degree angles, the nappe formed by continental collision. In places, the characteristic warping of the bands reveals the metamorphosis that took place under the immense pressure of colliding continents. And, within the rock, weird forms and figures emerged, such as curious eye-shaped patterns of quartz and what seemed like fossilized tree rings from a geological period that pre-existed trees as a species (Figs. 3.9, 3.10 and 3.11).

To be clear, we were quite certain we were not making the discovery of a century here; we came upon no Precambrian fossils. With little geological training among us, we lacked the expertise to speak of what might inform these features in the stone, or what they really were. We could, however, see how such features might provoke the imagination towards "folk geological" interpretations, and these curious considerations added a further dimension of weirdness to this unfamiliar place. At times, the peculiar elements resembled petrified living things embalmed into the rock. These aspects readily prodded the imagination in different directions, including reflections on what might lie beneath the surface of the rock and ground—and indeed beneath all of the readily observable world. What is important here is that the place, its landscape, these individual rocks are all infused with more-than-human stories and narratives that hint at unfamiliar and weird worlds. And, these stories, whether they are told or not, can be considered constituent parts of the objects that inspire them. Indeed, it has been said that "objects contain the possibility of all

3 THE "GLACIER" 67

Fig. 3.9 Weird banded gneiss of the nappe from 420+ million years ago, reaching for the sky

Fig. 3.10 A close-up of an eye pattern disturbingly oozing with quartz. Notice the fluidly twisted stratigraphy frozen in time

Fig. 3.11 Geological peculiarities inspired the attuned mind to come up with fantastic explanations, such as ancient and unknown lifeforms

situations" (Wittgenstein 1922). And, even a superficial survey of these potential stories embedded in the landscape reveals a wondrous potential for storytelling.

Yet, it was becoming more and more clear to us that it is not self-evident how stories of rock and of land can or should be told—that is, which stories are relevant or significant. Also, capturing and describing in words the emotional, affective and "otherworldly" qualities of the land is a complicated task. On more than one occasion, as we meandered about our peculiar rocks and rock formations in this desolate, desert-like landscape, our conversation turned to H. P. Lovecraft.

Lovecraft was an early twentieth-century American writer who became well known for his weird, science, fantasy and horror fiction and is most renowned for his creation of the Cthulhu Mythos. The Mythos, first devised by Lovecraft, but built upon by innumerous associated horror fiction writers since, is an imaginary pantheistic world (made such by Lovecraft's protégé August Derleth; Lovecraft himself did not envision a pantheon) that served in his tales as a backdrop element—a "pseudo-mythology", a term he himself used to describe the beings that appeared in his stories. It encompasses the shared elements, characters, settings and themes found in Lovecraft's works, but later became something of a canonical structure that enabled other writers to actively write about the myth-cycle or at least to allude to it in their stories. Lovecraft's fiction inspired "weird realism" (Harman 2012) and has latterly led to critical theory work on the horrors and monsters of the Anthropocene and on the vast gulf between cosmic, planetary horrors and meaningless human beings (Barad 2017; Tsing et al. 2017).

As we ambled about this landscape, its vastness, its uncanniness and the immensity of time which it represented twisted our perception and warped our sense of the place into what can only be described as Lovecraftian. Lovecraft weaved many of his stories around cosmological monsters that were too enormous, too weird and too ancient for the human mind to fathom. The space–time of his Cthulhu and other beings, whose temporalities span geological epochs and beyond, while being utterly incomprehensible to human beings, manages to diminish humanity to an insignificant speck of dust in the fabric of reality. Indeed, in his short essay on weird fiction, Lovecraft reflected that "the reason why time plays a great part in so many of my tales is that this element looms up in my mind as the most profoundly dramatic and grimly terrible thing in the universe. Conflict with time seems to me the most potent and fruitful theme in all human expression" (Lovecraft 1937). Indeed, the abyss of time is a foundation of all horror, and time can behave in deeply weird ways, as with the modernist anthropocentric illusion of time conceptualised as an arrow with the past and the future separated from our present reality (Fawver 2009; Olivier 2011; Carroll and Sperling 2020).

3.6 The Horror of Rock and the Emotions of Deep Time

The landscapes of Ritničohkka were primarily rockscapes composed of two main forms: solid bedrock and boulder fields. The landscape has been sculpted by ice and water, both during the Ice Age and the later reign of perennial ice and snow. This field, even in its heyday, was minuscule compared to true glaciers—not least the one that covered Fennoscandia during the Ice Age's glacial maximum, which was estimated to be over 2.4 kilometre thick, with a thickness of over a kilometre in the Háldi region. The presence of a glacier of such magnitude is difficult to even imagine, and it isn't any easier to envision the extractive forces by which such an entity would grind and pluck up the landscape and bedrock underneath it (see Patton et al. 2022).

Traces of the Ice Age are prominent in various forms throughout Fennoscandia today, but it was the *freshness*—the jagged bare rocks cleaved about randomly—of such traces on Ritničohkka that made the landscape here feel so special; the sense of the presence of deep time was unmistakable. Whether or not the ice field was as ancient in origin, the landscape it left behind echoed and conveyed a feeling of the post-Ice Age landscapes of some 11,000 years ago; it was as though we had been transported back to a Mesolithic world. This undoubtedly contributed to the sense of fantasy and weirdness that we experienced on Ritničohkka when surveying the slope where the ice field had been located just a few years prior.

The "fantastical" was present in the rockscape of Ritničohkka in both the guise of more or less striking "monumental" forms as well as the character of the exposed bedrock and the inclusions that the ice and other forces of erosion had exposed. The world beneath the surface of rock and the ground in general has fascinated people for centuries and millennia, undoubtedly inspiring diverse interpretations that exceed what today would be known as geological knowledge. Such notions are still present in folklore related to various aspects of rockscapes. The fossil-like inclusions that we encountered and wondered about on Ritničohkka are but one example of this.

Other types of inclusions in the Fennoscandian bedrock, the so-called crystal cavities, composed of high-quality quartz (such as rock crystal) and exposed by glacial forces, demonstrably attracted the interest of prehistoric people and quite likely were long associated with shamanistic ideas of a layered reality (Mökkönen et al. 2017). More generally, the subterranean has revealed all kinds of peculiar things, from fossils to ancient artefacts, which has presumably contributed to long-standing cultural ideas about the underground as an otherworldly reality of its own (see further e.g. Herva et al. 2022, 2024a, 2024b; Moshenska 2012).

While the rockscape that emerged from under the ice provoked a sense of prehistoric post-glacial landscapes, there is also a deeper temporal dimension to it. The bedrock of the Fennoscandian Shield, on which Finland almost exclusively rests, is staggeringly old, dating back some 2 billion years, whereas the bedrock in the Kilpisjärvi region is comparatively young, just some 500 million years old. The Kilpisjärvi bedrock represents an intrusion that makes this part of Finland geologically and environmentally unique, as it represents the tip of the Scandinavian mountain range that is primarily located in Sweden and Norway. Half a billion years is, of course, an age so old that it cannot really be properly conceived. This is around the time when multicellular life on Earth first emerged. Indeed, in 2013, trace fossils of early worms were discovered in a geological stratum near the Sána fjell in Kilpisjärvi.

Modest though these trace fossils may be, they nonetheless concretize and enable facing a very deep past and the otherness of ancient times that they represent. It is no coincidence that Lovecraft employs extremely ancient times as a tool for his horror. In his case, these take place not in the form of ordinary fossils but of the fossils (and sometimes the actual bodies) of "Ancient Ones" that built civilizations on Earth hundreds of millions of years ago and left behind vast and weird constructions, such as those described in his *At the Mountains of Madness* (1936; see also Hite 2021).

While Lovecraft's monsters are imaginary, it has been rationalized that the idea of monsters, such as dragons, was born of past peoples' struggle to make sense of the found fossils of extinct animals (e.g. Gould 2008 [1886]). While this may not suffice to explain monster imagery that has persisted in myriad forms from early Mesopotamia to the present day, such discoveries may indeed have mediated ideas of monsters across different places and times.

Whatever else they may be, "peculiar" or seemingly out-of-place rock formations and features in the rock (such as fossils and inclusions— embedded minerals, water, gas or other rocks) have been part of human landscapes throughout time and have undoubtedly sparked curiosity and meaning-making, likely with associations with ancient times (regardless of how exactly they may have been conceived in different cultures), as rock "naturally" elicits ideas, notions and connotations of ancient times and eternity (e.g. Parker Pearson and Ramilisonina 1998). The folklore most typically associated with rocks and stones readily demonstrates both the cultural interest in geological processes and formations and a tangible presence of weirdness and the monstrous in human lived environments. In Finnish folklore, for instance, rocks and rocky places are commonly associated with (ancient) non-humans, such as trolls and giants.

Practically all these rockscapes were carved by glacial ice, a reality that imbues all ice with a sense of awe. That bits and pieces of this ancient bedrock could have simply been plucked out by the omnipresent glacier of the last Ice Age—extremely recently in geologic time but long, long ago in human time—is at once mesmerizing and terrifying. The ice sheet, a massive "Cyclopean" (an adjective Lovecraft was fond of) monster of immense transformative power, still looms amidst the mise-en-scène of north-west Lapland, whose present-day geography was made by the glaciofluvial action of the glacier's heaving and death. With minimal imagination, one can see the entire landscape as the final imprint of the decaying corpse of the vast Fennoscandian Ice Sheet, which even after its demise overshadows anything humanity has ever constructed. A true cosmic monstrosity that blurs the lines between the real-world and weird fiction (Fig. 3.12).

One might consider such musings to fall into the fanciful category of geomythology, the power of which is evident worldwide, give or take cultural differences (Piccardi and Masse 2007). The depiction of the Fennoscandian Ice Sheet as a cosmic monster, in fact, echoes an actual mythological reference: that of Ymir, the primordial frost giant, from whose slain corpse the northern world was created, as recounted in the Old Norse *Poetic Edda* (see Pettit 2023):

> From Ymir's flesh the earth was formed,
> and from his bones boulders,

Fig. 3.12 Markus getting unsuspectingly crushed by Ymir's ghost

the sky from the skull of the frost-cold giant,
and from his "sweat" the sea.

This early poem resonates with perceptions in indigenous cosmologies of glaciers as animate and conscious entities—beings rather than things. Such long-standing indigenous worldviews have made their way into contemporary associations in a number of European and North American contexts, bringing us back to the lamenting and mourning of dying glaciers as a response to the horrors of the Anthropocene (Varnajot and Saarinen 2022, and references therein to glaciers as person-like beings in different contexts; also Árnason and Hafsteinsson 2020).

The origin stories of myths are most often untraceable in practice. Serious attempts usually fall apart under critical review and are taken up by pseudo-scientific pursuits with the potential to grab headlines and feature in docu-series. Rarely do they amount to any substantial revelation about the myths themselves. Still, a science-based approach is represented by the concept of the geomyth. Introducing the concept in 1968, geologist Dorothy Vitaliano defined it as seeking "to find the real geologic event underlying a myth or legend to which it has given rise; thus [helping to] convert mythology back into history" (Vitaliano 1968).

However, as we discovered in the strange configurations of rocks and their forms, these geomyths do not have to represent a traceable geologic event, a catastrophic apocalypse directly experienced by humanity. Actually applying a geomyth framing to deep, more-than-human time may, counterintuitively, prove more scientific than tracing specific events. Instead of working to establish links that might remain forever unproven, identifying mundane processes and gradual time spans that are untraceable to a single source event can provide bounteous sources for imagination and myth making.

Deep time is a potential source of existentialist horror because the vastness of geological and cosmic time, spanning millions and billions of years, renders not just human life but all life on Earth utterly insignificant and meaningless, much in the manner of Lovecraft's cosmic monsters and horrors (see Carroll and Sperling 2020) (Fig. 3.13). Apart from this, time can behave weirdly, as illustrated, for instance, by experiences of haunting and phenomena such as déjà-vu. Indeed, the roots of the very word "weird" refer to twisted and looped (Morton 2016: 5–6). The rocks on which we walked atop the boulder fields of Ritničohkka represented hundreds of millions of years, compressed and flattened surfaces comprising frozen vortexes of time. Such materialization of deep time can cause vertigo—kinaesthetically and mentally. For us, this vertigo became embedded in the landscape of Ritničohkka, which in itself amplifies a sense of weird temporalities in the form of the sense of timelessness and eternity.

"Weird temporalities", for Carroll and Sperling (2020: 9, with reference to Harman 2012: 81), "reveal and render strange the properties of normative time". And, indeed, time in the High North is readily strange in the sense that, as has long been marvelled at in historical travelogues, nights are nightless during the summer and days dayless in the winter. In this simple way, the assumed normal order of things and temporalities of the Eurocentric world are compromised, inverted. More generally, "weird time" is something that "disobeys expectations of temporal qualities such as sequence, span, speed, synchronization, rhythm, orientation, causality, coherence, or intensity" (Carroll and Sperling 2020: 9).

Not only our sense of space, motion and proportion was being actively worked on by the mountains and the trickling meltwater running under the rocks at our feet, but so was our sense of time. As deep time opened its doors, the present became strangely urgent. And the present, and our

Fig. 3.13 Stepping onto the abyss of time. A recently exposed subglacial surface, with twisted mineral sheets characteristic of gneiss, reveals mesmerizing metamorphoses dating from the Silurian Age, flatting the vastness of time into a surface

presence in it, began to resemble an intrusion, difficult to fully grasp or describe. As if our being there witnessing the moment was unwelcomed by the timeless beings incubating in a blended haze of landscape and imagination.

REFERENCES

Anderson, D.G. 2000. *Identity and Ecology in Arctic Siberia: The Number One Reindeer Brigade.* New York: Oxford University Press.
Árnason, A., and S.B. Hafsteinsson. 2020. A Funeral for a Glacier: Mourning the More-Than-Human on the Edges of Modernity. *Thanatos* 9 (2): 46–71.
Barad, K. 2017. No Small Matter: Mushroom Clouds, Ecologies of Nothingness, and Strange Topologies of Spacetimemattering. In *Arts of Living on a Damaged Planet: Ghosts and Monsters of the Anthropocene*, ed. A. Tsing, H. Swanson, E. Gan, and N. Bubandt, 103–121. Minneapolis: University of Minnesota Press.
Buhl, A. 2023. The Death of Clark Glacier: Practices of Mourning and the Possibility of Post-Anthropogenic Memory. *Leviathan: Interdisciplinary Journal in English* 9. https://doi.org/10.7146/lev92023136281.
Carey, M. 2007. The History of Ice: How Glaciers Became an Endangered Species. *Environmental History* 12 (3): 497–527.
Carey, M., M. Jackson, A. Antonello, and J. Rushing. 2016. Glaciers, Gender, and Science: A Feminist Glaciology Framework for Global Environmental Change Research. *Progress in Human Geography* 40 (6): 770–793. https://doi.org/10.1177/0309132515623368.
Carroll, J.S., and A. Sperling. 2020. Weird Temporalities: An Introduction. *Studies in the Fantastic* 9: 1–22.
Cruikshank, J. 2005. *Do Glaciers Listen? Local Knowledge, Colonial Encounters, and Social Imagination.* Vancouver: University of British Columbia Press.
Fawver, K. 2009. "Present"-ly Safe: The Anthropocentrism of Time in H. P. Lovecraft's Fiction. *Journal of the Fantastic in the Arts* 20 (2): 248–261.
Fjellström, M., O. Seitsonen, and H. Wallén. 2022. Mobility in Early Reindeer Herding. In *Domestication in Action. The Anthropology and Archaeology of Reindeer Domestication in Fennoscandia*, 187–212. Cham: Palgrave Macmillan.
Gould, C. 2008 [1886]. *Mythical Monsters.* New York: Cosimo.
Graves-Brown, P., and J. Schofield. 2020. Encountering Landscape: Travel as Method. *Landscapes* 20 (1): 61–84.
Hakonen, A. 2021. Local Communities of the Bothnian Arc in a Prehistoric World. PhD thesis, University of Oulu.
Harman, G. 2012. *Weird Realism: Lovecraft and Philosophy.* Winchester: Zero Books.

Harris, P.A. 2015. Introduction: David Mitchell in the Labyrinth of Time. *SubStance* 44 (1): 3–7.

Herva, V.-P., and A. Lahelma. 2020. *Northern Archaeology and Cosmology: A Relational View*. Abingdon: Routledge.

Herva, V.-P., and O. Seitsonen. 2020. The Haunting and Blessing of Kankiniemi: Coping with the Ghosts of the Second World War in Northernmost Finland. In *Entangled Beliefs and Rituals*, ed. T. Äikäs and S. Lipkin, 225–235. Helsinki: The Archaeological Society of Finland.

Herva, V.-P., and T. Ylimaunu. 2009. Folk Beliefs, Special Deposits, and Engagement with the Environment in Early Modern Northern Finland. *Journal of Anthropological Archaeology* 28 (2): 234–243.

Herva, V.-P., A. Varnajot, and A. Pashkevich. 2020. Bad Santa: Cultural Heritage, Mystification of the Arctic, and Tourism as an Extractive Industry. *The Polar Journal* 10 (2): 375–396.

Herva, V.-P., T. Komu, and T. Paphitis. 2022. Extraordinary Underground: Fear, Fantasy and Future Extraction. In *Resource Extraction and Arctic Communities: The New Extractivist Paradigm*, ed. S. Sörlin, 166–182. Cambridge: Cambridge University Press.

Herva, V.-P., G. Moshenska, T. Paphitis, T. Äikäs, I. Banks, R. Nurmi, O. Seitsonen, and S. Thomas. 2024a. Weird Quantities: Characterising Monstrous Landscapes of Extraction in the Anthropocene. *Time and Mind*: 1–24 (online first). https://doi.org/10.1080/1751696X.2024.2353748.

Herva, V.-P., O. Seitsonen, T. Paphitis, T. Komu, G. Moshenska, and R. Nurmi. 2024b. Spiralling into a Labyrinth of Cultural Fantasies and Extractivism: Treasures, Extraordinary Undergrounds, and the 'Temple of Lemminkäinen' (Sipoo, Finland). In *Connecting with Ambivalent Heritage: Creative Uses of Postindustrial Spaces*, ed. T. Äikäs and T. Matila, 131–154. London: Bloomsbury.

Hillier, J. 2017. Lines of Becoming. In *The Routledge Handbook of Planning Theory*, ed. M. Gunder, A. Madanipour, and V. Watson. Abingdon: Routledge.

Hite, K. 2021. *Tour de Lovecraft: The Destinations*. Alexandria, VA: Atomic Overmind Press.

Hirvas, H., P. Lintinen, and P. Kosloff. 2000. An Extensive Permanent Snowfield and the Possible Occurrence of Permafrost in Till in the Ridnisohkka Area, Finnish Lapland. *Bulletin of the Geological Society of Finland* 72 (1): 47–56.

Hirvas, H., P. Lintinen, A.E.K. Ojala, and H. Vanhala. 2005. Geological Characteristics of the Halti-Ridnitšohkka Region, Enontekiö, Finland. In *Quaternary Studies in the Northern and Arctic Regions of Finland*, ed. A.E.K. Ojala, 7–12. Helsinki: Geological Survey of Finland.

Howe, C., and Boyer, D. 2024. "The Okjökull Memorial and Geohuman Relations". *Social Anthropology* 32 (1): 30–45.

Itkonen, T.I. 1948. *Suomen lappalaiset*. Helsinki: WSOY.

Jackson, M. 2015. Glaciers and Climate Change: Narratives of Ruined Futures. *Wires Climate Change* 6: 479–492. https://doi.org/10.1002/wcc.351.

Joks, S., L. Østmo, and J. Law. 2020. Verbing meahcci: Living Sámi Lands. *The Sociological Review* 68 (2): 305–321.

Kaplan, L. 2003. Where the Paranoid Meets the Paranormal: Speculations on Spirit Photography. *The Art Journal* 62 (3): 18–29.

Karlsson, N. 2006. Bosättning och resursutnyttjande. Miljöarkeologiska studier av boplatser med härder inom perioden 600–1900 e. Kr inom skogssamiskt område. PhD dissertation. Umeå, Umeå University.

Kauhanen, J. 2024. *Terra nullius ja paikallisyhteisöt modernisaation puristuksessa 1950- ja 1960-luvun Lapin vesivoimarakentamisessa Pohjoiskalotin suunnitelmaksi jäänyt rakentamishanke ja Sodankylän säännöstelyaltaat*. Rovaniemi: Lapin yliopisto.

Kenny, S. 2021. "The Story of Okjökull—The Icelandic glacier that disappeared". Muchbetteradventures.com, 8 November. Available at: https://www.muchbetteradventures.com/magazine/okjokull-glacier-history-iceland-disappeared.

Kohonen, J., and O.T. Rämö. 2005. Sedimentary Rocks, Diabases, and Late Cratonic Evolution. In *Precambrian Geology of Finland: Key to the Evolution of the Fennoscandian Shield*, ed. M. Lehtinen, P.A. Nurmi, and O. Rämö, 563–603. Amsterdam: Elsevier.

Lahtinen, S., L. Jivall, P. Häkli and M. Nordman. 2022. Updated GNSS velocity solution in the Nordic and Baltic countries with a semi-automatic offset detection method. *GPS Solutions* 26 (1): 9. https://doi.org/10.1007/s10291-021-01194-z

Länsman, A.-S. 2004. *Väärtisuhteet Lapin matkailussa Kulttuurianalyysi suomalaisten ja saamelaisten kohtaamisesta*. Inari: Kustannus-Puntsi.

Lehtovaara, J.J. 1995. *Kilpisjärven ja Haltin kartta-alueiden kallioperä. Explanation to the Maps of Pre-Quaternary Rocks*. Espoo: Geological Survey of Finland.

Lovecraft, H.P. 1937. Notes on Writing Weird Fiction [essay]. http://www.hplovecraft.com/writings/texts/essays/nwwf.aspx.

Luukas, J., and J. Kohonen. 2021. Major Thrusts and Thrust-Bounded Geological Units in Finland: A Tectonostratigraphic Approach. *Geological Survey of Finland, Bulletin* 412: 81–114.

Magga, P., and S. Terваniemi. 2018. Belonging to Sápmi—The Sámi Conceptions of Home and Home Region. In *Knowing from the Indigenous North. Sámi Approaches to History, Politics and Belonging*, ed. T.H. Eriksen, S. Valkonen, and J. Valkonen, 75–90. Routledge: Abingdon.

Majka, J., Y. Be'eri-Shlevin, D.G. Gee, A. Ladenberger, S. Claesson, P. Konečný, and I. Klonowska. 2012. Multiple Monazite Growth in the Åreskutan

Migmatite: Evidence for a Polymetamorphic Late Ordovician to Late Silurian Evolution in the Seve Nappe Complex of West–Central Jämtland, Sweden. *Journal of Geosciences* 57 (1): 3–23.

Mazzullo, N., and T. Ingold. 2008. Being Along: Place, Time and Movement Among Sámi People. In *Mobility and Place. Enacting Northern European Peripheries*, ed. J.O. Baerenholdt and B. Granås, 27–38. Aldershot: Ashgate.

Mökkönen, T., K. Nordqvist, and V.-P. Herva. 2017. Beneath the Surface of the World: High-Quality Quartzes, Crystal Cavities, and Neolithization in Circumpolar Europe. *Arctic Anthropology* 54 (2): 94–110. https://doi.org/10.3368/aa.54.2.94.

Morton, T. 2016. *Dark Ecology: For a Logic of Future Coexistence*. New York: Columbia University Press.

Moshenska, G. 2012. M.R. James and the Archaeological Uncanny. *Antiquity* 86 (334): 1192–1201. https://doi.org/10.1017/S0003598X00048341.

Näkkäläjärvi, K. 2013. Jauristunturin poropaimentolaisuus: kulttuurin kehitys ja tietojärjestelmä vuosina 1930–1995. Ph.D. dissertation, University of Oulu, Oulu.

Olivier, L. 2011. *The Dark Abyss of Time: Archaeology and Memory*. Lanham: AltaMira Press.

Orlove, B., E. Wiegandt, and B.H. Luckman, eds. 2008. *Darkening Peaks. Glacier Retreat, Science, and Society*. Berkeley and Los Angeles: University of California Press.

Parker Pearson, M., and Ramilisonina. 1998. Stonehenge for the Ancestors: The Stones Pass on the Message. *Antiquity* 72 (276): 308–326. https://doi.org/10.1017/S0003598X00086592.

Patton, H., A. Hubbard, K. Andreassen, M. Winsborrow, and A.P. Stroeven. 2016. The Build-Up, Configuration, and Dynamical Sensitivity of the Eurasian Ice-Sheet Complex to Late Weichselian Climatic and Oceanic Forcing. *Quaternary Science Reviews* 153 (1): 97–121.

Patton, H., A. Hubbard, J. Heyman, et al. 2022. The Extreme Yet Transient Nature of Glacial Erosion. *Nature Communications* 13: 7377. https://doi.org/10.1038/s41467-022-35072-0.

Pettit, E. 2023. *The Poetic Edda: A Dual-Language Edition*. Cambridge: Open Book Publishers.

Piccardi, L., and W.B. Masse, eds. 2007. *Myth and Geology*. London: Geological Society.

Pilø, L., E. Finstad, C.B. Ramsey, J.R.P. Martinsen, A. Nesje, B. Solli, V. Wangen, M. Callanan, and J.H. Barrett. 2018. The Chronology of Reindeer Hunting on Norway's Highest Ice Patches. *Royal Society Open Science* 5: 171738. https://doi.org/10.1098/rsos.171738.

Pollack, H. 2009. *A World Without Ice*. New York: Avery.

Provant, Z., and Carey, M. 2022. "Who is Killing the Glaciers? From Glacier Funerals to Glacier Autopsies" Edge Effects, 3 November. Available at: https://edgeeffects.net/glacier-funerals.

Ruotsala, H. 2002. *Muuttuvat palkiset: elo, työ ja ympäristö Kittilän Kyrön paliskunnassa ja Kuolan Luujärven poronhoitokollektiiveissa vuosina 1930–1995*. Helsinki: Suomen muinaismuistoyhdistys.

Salmi, A.-K., and O. Seitsonen. 2022. Effects of Reindeer Domestication on Society and Religion. In *Domestication in Action: The Anthropology and Archaeology of Reindeer Domestication in Fennoscandia*, ed. A.-K. Salmi, 215–248. Cham: Palgrave Macmillan.

Schanche, A. 2002. Meahcci – den samiske utmark. *Diedut* 1: 156–171.

Seitsonen, O., and N. Égüez. 2022. Here Be Reindeer: Geoarchaeological Approaches to the Transspecies Lifeworlds of the Sámi Reindeer Herder Camps on the Tundra. In *Materiality and Objects*, ed. P. Halinen and J. Taivainen, 124–145. Helsinki: The Finnish Antiquarian Society.

Seitsonen, O., and M. Fjellström. 2022. Habitation Sites and Herding Landscapes. In *Domestication in Action: The Anthropology and Archaeology of Reindeer Domestication in Fennoscandia*, ed. A.-K. Salmi, 153–186. Cham: Palgrave Macmillan.

Seitsonen, O., and E. Koskinen-Koivisto. 2017. 'Where the F… Is Vuotso?': Heritage of Second World War Forced Movement and Destruction in a Sámi Reindeer Herding Community in Finnish Lapland. *International Journal of Heritage Studies* 24 (4): 421–441. https://doi.org/10.1080/13527258.2017.1378903.

Seitsonen, O., and S. Viljanmaa. 2021. Landscapes of Sámi Reindeer Domestication and Pastoralism in the Gilbbesjávri Region, Sápmi, Northernmost Europe ca. 700–1800 A.D. *Journal of Field Archaeology* 46 (3): 172–191.

Seitsonen, O., M. Fjellstöm, A. Hakonen, V.-P. Herva, R. Norum, S. Seitsonen, S. Seitsonen, E. Seitsonen, and E. Seitsonen. forthcoming. Archaeology of Finland's Only 'Permanent Glacier': The Past, Present, and Future of the Ritničohkka Snowfield, Sápmi. *Journal of Glacial Archaeology* (online first).

Selley, R.C., L.R.M. Cocks, and I.R. Plimer, eds. 2005. *Encyclopedia of Geology*, vol. 2. London: Elsevier.

Sipilä, P. 1992. The Caledonian Halti-Ridnitsohkka Igneous Complex in Lapland. *Geological Survey of Finland, Bulletin* 362.

Sörlin, S. 2015. Cryo-History: Narratives of Ice and the Emerging Arctic Humanities. In *The New Arctic*, ed. B. Evengård, J. Nymand Larsen, and Ø. Paasche. Cham: Springer https://doi.org/10.1007/978-3-319-17602-4_24.

Taylor, W., J.K. Clark, B. Reichhardt, G.W.L. Hodgins, J. Bayarsaikhan, O. Batchuluun, J. Whitworth, M. Nansalmaa, C.M. Lee, and E.J. Dixon. 2019. Investigating Reindeer Pastoralism and Exploitation of High Mountain Zones in Northern Mongolia Through Ice Patch Archaeology. *PLoS ONE* 14 (11): e0224741. https://doi.org/10.1371/journal.pone.0224741.

Tsing, A., H. Swanson, E. Gan, and N. Bubandt, eds. 2017. *Arts of Living on a Damaged Planet: Ghosts and Monsters of the Anthropocene*. Minneapolis: University of Minnesota Press. http://www.jstor.org/stable/10.5749/j.ctt1qft070

Vanhala, H., I. Suppala, P. Lintinen, H. Hirvas, and A.E.K. Ojala. 2005. Application of Electrical and Electromagnetic Methods in Studying Frozen Ground and Bedrock—Results from Ridnitsohkka, Northern Finland. In *Quaternary Studied in the Northern and Arctic Regions of Finland*, ed. A.E.K. Ojala, 13–22. Espoo: Geological Survey of Finland.

Varnajot, A., and J. Saarinen. 2022. Emerging Post-Arctic Tourism in the Age of Anthropocene: Case Finnish Lapland. *Scandinavian Journal of Hospitality and Tourism* 22 (4–5): 357–371.

Vitaliano, D.B. 1968. Geomythology: The Impact of Geologic Events on History and Legend with Special Reference to Atlantis. *Journal of the Folklore Institute* 5 (1): 5–30. https://doi.org/10.2307/3813842.

Wittgenstein, L. 1922. *Tractatus Logico-Philosophicus*. New York: Harcourt, Brace & Company Inc.

Open Access This chapter is licensed under the terms of the Creative Commons Attribution 4.0 International License (http://creativecommons.org/licenses/by/4.0/), which permits use, sharing, adaptation, distribution and reproduction in any medium or format, as long as you give appropriate credit to the original author(s) and the source, provide a link to the Creative Commons license and indicate if changes were made.

The images or other third party material in this chapter are included in the chapter's Creative Commons license, unless indicated otherwise in a credit line to the material. If material is not included in the chapter's Creative Commons license and your intended use is not permitted by statutory regulation or exceeds the permitted use, you will need to obtain permission directly from the copyright holder.

CHAPTER 4

Mountain Beings

> Bones project time
> Our scaffolds,
> us and the reindeer,
> picked clean and bleached,
> turn into archives, museums
> Encapsulated in ice they thrive
> spat out by the melt they die
> incineration in slow motion
> Which would my tibia prefer?
> carried here by the carnivore
> To pass messages, violent, from the time deep
> or to pass and say nothing at all?
> —"*Poe*tic Pastiche" #1, Aki Hakonen

Traversing the landscape continued to be tedious and slow, with loose rocks rolling underfoot. Strange rock formations raised existentialist questions, giving rise to a general sense that the environment was "dead" and simultaneously vibrating with more-than-human life and consciousness. This contradiction felt like a significant characteristic of the land as it presented itself to us. But how can one "know" and describe such a landscape? Monotonous and exciting at the same time, the landscape was shapeless yet full of shapes, some familiar and others bizarre. It evoked a feeling of strangeness of its own, partly because of how we engaged with it and partly because of our imaginaries and projections of the place. In any case, something in this physical and perceptual landscape resonated conceptually with the monsters of the Anthropocene and the monstrous. The place was simultaneously too real and too ethereal, nondescript and fantastic. Indeed, oppositions and contradictions of all kinds have always been central to perceptions and descriptions of Lapland and the High North, a land of gloom and plenty, of light and darkness, of visions, ghostly presences, mirages and illusions (Davidson 2005; Naum 2016; Herva et al. 2020).

4.1 Nearing the Monstrous, Weirding Description

The question arises: how can one describe weird monstrous entities in words? On the one hand, they are by definition something that goes beyond verbal description, and yet we should somehow try to "near" unfamiliar worlds (Olsen and Pétursdóttir 2021)—that is, connect with the unfamiliar on a verbal level. Harman analyses in depth the literary techniques that Lovecraft employed (Harman 2012). For instance, for an "extravagant imagination", Cthulhu yields "simultaneous pictures of an octopus, a dragon, and a human caricature", which is not "unfaithful to the spirit of the thing", but—critically—Cthulhu is not simply a composite of these three things, but rather "the general outline of the whole" is something more and something different (Harman 2012: 57–59). This can be related to Lovecraft's technique of using "the word 'or' in order to treat two divergent realities as though they were familiar next-door neighbours in the continuum of being" (Harman 2012: 91). Although the way we felt about and sensed the place and geology of Ritničohkka was not exactly "Cthulhuesque", there was something decidedly Lovecraftian about them, along the lines of the haunting "ontological indeterminacy of time-being/being-time in its materiality" (Barad 2017: 113).

There were, for instance, blocks of grey rock with smooth surfaces, like gigantic ice cubes that seemed as if dropped on the boulder field. There were also other curious formations that ice had carved on the side of the fjell, such as canals in the bedrock that, in some cases, were interrupted by protruding out-of-place blocks of rock (Fig. 4.1). Non-Euclidean geometries. In a word, Lovecraftian—that is, not just uncanny but more deeply and disturbingly weird, as if space and time and geographies were skewed in a profound yet perceptually subtle manner. There was a feeling,

Fig. 4.1 Things (and people) out of place

anchored in different landscape features, that something almost-but-not-quite-perceivable weirdness rested beneath this curious landscape. A sense of invisible and more-than-human consciousness. While it was not exactly R'lyeh, the submerged stone city where Cthulhu lies dreaming and whose geometry is said to be "all wrong", Ritničohkka still had an undefinable sense of "moreness" and out-of-placeness with respect to the land, which at quick glance might appear merely barren and desolate (Fig. 4.2).

4.2 A Sea of Stones

It only occurred to us later that one way of approaching this mountain landscape and the feelings it invoked is to conceive of it as a kind of a sea. The germ for this idea was originally planted when looking at the wavy, undulating landscape from the helicopter. The story or deep history of this land and how it has formed over the last thousands and thousands of years is very much a story of sea and ice. The ice sheet that covered Fennoscandia and northern Europe during the last Ice Age, the Weichselian glaciation, shaped this landscape, together with the melting of the ice sheet and subsequent post-glacial processes, leaving behind this seemingly endless boulder field. Perceptually it is a stone sea, an imprint of water and ice on the now terrestrial landscape, a residual or fossilized

Fig. 4.2 Cliff walls erected by geological forces present imposing surfaces with indescribable, "wrong" geometries. A melting ice patch lies mournfully at the feet of the high escarpment that rises abruptly for over 20 unscalable metres

seascape. Indeed, a metaphysical connection between stone and water has been conceived in some cultures, as in the case of marble and limestone in classical and Christian mindscapes (Barry 2007).

The usefulness of the sea metaphor in this context hinges primarily on widespread cultural associations of the sea as a non-human domain of otherness and otherworldliness. The sea is ambiguous and amorphous, seemingly timeless and yet constantly changing, poorly known, and hiding real and imagined mysteries in its depths (Cunliffe 2017; for historical and ethnographic examples and expressions of this from the Baltic Sea region, see Westerdahl 2005). Some of these qualities resonate, perceptually and cognitively, with Ritničohkka and its surroundings as a stone sea, monotonous and shapeless but, on closer look, like waves in the sea, full of shapes and constant variation, albeit seemingly frozen and motionless. Thus, it is not only a sea but a *very weird sea*. We are not the first to observe this parallel between fjell landscapes and the sea. Describing the remote fjells of the Petsamo area in eastern Lapland, the Finnish novelist Pentti Haanpää mused how "the endless, solid waves of the fjells depress and burden. A living image […] of the length of eternity" (Haanpää 1980: 97).

The glacial forces of ice and water have shaped the northern Fennoscandian landscapes and, in a sense, imprinted themselves on it, as discussed above, but for our purposes the idea of the rocky tundra as a seascape is useful because it enables grasping some of the perceptual and emotive qualities of boulder fields. The sea is a weird world, and a sea of boulders is doubly weird, something that goes beyond the ordinary and fathomable weirdness of the sea—not merely into otherness but into the unreal and out of this world. There's a dream-like essence to the fjell landscape, not only when seen from the helicopter but also on the ground. In this instance, to us, it was enchanting but also horror-laden in an existentialist sense that goes beyond the mere physical threats that a remote northern land might pose.

On the first day of our fieldwork, these feelings—like the landscape around us—were vague and formless, but they haunted several of us throughout our stay up on the mountain-sea and became entangled with more specific experiences. These involved, for instance, illusion or vision-like experiences, such as when certain views opening from Ritničohkka felt like looking at some Near Eastern desert landscapes (see Herva 2021).

The monstrousness we felt in the landscape was not strictly malicious. Instead, the feeling arose from the scale and the myriad uncertainties of our experienced space–time and of being in that environment. It transcended the this-worldly and the otherworldly and was ambivalent and indefinite in ways that are difficult to verbalize (see e.g. Harman 2012). The familiarity we had with high northern landscapes and worlds was disturbed by our attuning to the Ritničohkka landscape, where a multitude of out-of-ordinary features and affective qualities assailed us. Characterizations, such as the stone sea, do not precisely define the landscape—just as Cthulhu is not merely a combination of an octopus, a dragon and a human caricature—but suffice to reflect a fleeting glimpse of its strange nature.

But whatever else Ritničohkka is, it is undeniably a fjell, and fjells are effectively old, rounded mountains. Mountains, like glaciers, are imposing, awe-inspiring and potentially terrifying landscape elements to which many cultures assign otherworldly dimensions and qualities. This also applies to fjells, which have a haunting and mysterious presence in Lapland landscapes, as described by travellers and ethnographers, albeit often in a romanticized manner and with poetic licence (e.g. Paulaharju 1927: 4–11). There are stories of different origins about fjells as having been—and perhaps still residually being—sentient and conscious, more-than-human beings frozen in time (Valtonen 2019; for various examples of mountains perceived as having special properties, see Bernbaum 2022). Such narratives are commonplace across the border in mountainous Norway (Petersen 2017) and are reflected, for instance, in contemporary popular culture by Norwegian films such as *Troll Hunter* (2010) and *Troll* (2022).

A vague, lingering sense of non-human consciousness and awareness emanating from the environment was a part of our experience of Ritničohkka, likely generated by myriad cues in the landscape that resonated with things in our heads, whether from reading Lovecraft, Nordic folklore, Sámi ethnographies or a host of other sources deviating from the modernist natural science model. We are not alone in having such notions. Mountains—much like the sea—are strange entities with vague boundaries that transgress modernist categories; they are neither living nor non-living, animate nor inanimate, moving nor non-moving, and create a sense of stability while simultaneously hinting at an inhuman timescale of change (Petersen 2017). It is perhaps no surprise in this view that the landscape and wilderness themselves can function as monsters, featuring in this role in films such as *Dead Snow* (2009) and in Scandinavian horror fiction (Piatti-Farnell 2021). Horror and monstrousness are embedded in the environment, which storytellers may use to great effect "[r]ather than inventing a monster with an arbitrary number of tentacles and dangerous sucker-mouths and telepathic brains" (Harman 2012: 22). This is also one reason why Ritničohkka affected us so powerfully: the "monster" was there, everywhere; invisible and undetectable.

4.3 Mountain Beings and Mineral Evolution

In his article, *At the mountains of monstrosity* (inspired by Lovecraft's novella *At the Mountains of Madness*), Daniel Petersen (2017) discusses human affinity with mountains, and mountains as gigantic monstrous bodies. Petersen reflects on how, upon visiting Lysefjord in Norway, he "was struck with something of a corresponding sense that the mountains towering over us on our fjord cruise were the magnificent, cyclopean heads of Lovecraftian Old Ones rearing up with aching slowness out of this enormous glacial scar in the earth" (p. 72). He goes on to map various similarities between mountains and monsters, including not just their scale and morphology, but also indefinite boundaries and inhuman temporalities: mountains as entities are "strange, hybrid, ghostly"—they "transgress[…] our binaries of life/nonlife, animate/inanimate" and "blur[…] movement and non-movement as well" (Petersen 2017: 76). As for the Kilpisjärvi region, we might also recall that Sána and Malla have been envisioned—even if only for the entertainment of tourists—as petrified ancients, while an info blurb on the Sána Science Trails tells us that biocrust is the "living *skin* of the fjell" (emphasis added).

Human engagement with mountains, Petersen (2017) argues, makes the boundaries between the living and non-living permeable, porous, uncertain and transcendent—even obsolete—with people joining the mountains as "terrestrial aliens", the smaller swallowed by the larger. For Petersen, humans and the lithosphere "occupy one another" as "[w]e can only see the earth as it sees itself by remembering we are the earth, beings formed of mud and rock and water who never escape those elements but ever return to and merge with them" (p. 84). Consequently, within such landscapes, we humans are not only made of our bones, flesh and skin but also of non-human mountain matter and "of steep rocky walls, moss, streams, scrub, wind, rain, ice" (p. 84). While we may not have experienced this in quite as graphic and dramatic terms on Ritničohkka, we certainly did feel, on occasion, our boundaries dissolving in relation to the landscape, being part of and merging with that ancient mountain with its weirder features and dimensions. Ritničohkka engaged with us—most readily through rocks, ice, wind and clouds—as much as we engaged with Ritničohkka. We were part of a mountain being (Fig. 4.3).

In many cultures, rock itself is—or can be—alive or living matter (e.g. Breda 2022). The peculiar inclusions in rock, such as the slightly disturbing eye-shaped, quartz-oozing inclusion in Figure 3.10, on Ritničohkka were a reminder of this. They were reminiscent of Lovecraft's barrel-shaped, star-headed beings discovered "at the mountains of madness" in the Antarctic. While we didn't identify any real fossils, their presence on Ritničohkka is not out of the question. As mentioned before, actual fossils—specifically trace fossils—were recently found on the fjell Sána in Kilpisjärvi. These trace fossils of worm-like beings represent some of the earliest multicellular forms of life that developed rapidly some 500 million years ago in the Cambrian period. Fossils comprise a tangible link between the present and the unimaginably deep past with its (from a human perspective) weird beings and their lifeworlds.

But there is more to this entanglement of mineral and organic worlds. Hazen et al. (2009) suggested the idea of "mineral evolution", which

Fig. 4.3 Mountain beings. Roger and Vesa-Pekka pondering on a weird landscape

proposes that minerals and organic life have co-evolved, with changes in the mineral world creating conditions for the emergence and development of organic life, and organic life, in turn, contributing to the generation of new minerals. This provides an additional dimension to the notion of mountain beings and the connectedness and reciprocity between humans and the lithosphere. It has been noted that lichen, a primordial lifeform abundant in the Northern Arctic, specifically is responsible for originally

Fig. 4.4 Minerals evolving, with many forms of lichen extracting minerals from the rock and renewing the world. Moss and lichen are slowly creeping towards a freshly bleached bone

creating the fertile soils of Earth by feasting on the minerals of otherwise impenetrable stone and dying in accumulating strata. This genesis, a mineral evolution, was continuing to occur before our eyes, as if time itself had rewound back to the beginning of life on land (Fig. 4.4).

Speaking of a genesis, the Christian creation myth also explicates a close, culturally conceived association between humans and the mineral domain. After all, the first human was made out of clay, into which God breathed life and spirit. The notion of breathing, as associated with the matter and "substance" of life, is interesting here, as wind can also be considered a sentient, spiritual and animate being (e.g. Ingold 2000: 48–49, 91–93, 102), and it ties together the different human and non-human constituents that make up mountain beings (see also Sect. 6.6) (Fig. 4.4).

4.4 Absences and Presences of Life, Death and Sentience

Arctic environments are harsh and may appear quite lifeless, yet the Arctic is also characterized, biologically and culturally, by diversity and the omnipresence of life force (Ingold 2019). The interplay between life and death is very prominently present in the High North and its networks of relations between diverse entities and beings (Herva et al. 2024). At the same time, the combination of monotony and diversity is what makes high northern landscapes special and fascinating. There is a peculiar uncertainty to the presence and forms of life here. In the human-experienced world of the High North, conscious and sentient presences are here, there and everywhere—even when actual sightings of life are elusive. In other words, things are not necessarily what they seem, which is an integral aspect of relational (or shamanistic-animistic) lived realities, in which everything has the potential to constantly move and change.

Life in the Arctic is largely hidden, which is particularly clear in settings such as the boulder fields of Ritničohkka. There are no trees or prominent vegetation—and, indeed, there is also a lack of noticeable animal life. The free-ranging reindeer wander there only seasonally, and now that the snow patches are gone, they may not find the pasture as attractive as before. By and large, life is hiding in minuscule forms and expressions. And therein lies another dissonance: life is everywhere yet barely visible. This dissonance strikes a chord with Algernon Blackwood's classic horror story, *The Willows* (1907), in which a monotonous willow-dominated riparian landscape of the Danube oozes a sense of more-than-human consciousness and sentience that, as the characters sense it, is hidden behind a thin veil between the this-worldly and the otherworldly.

Much of biological life on Ritničohkka is unnoticeable, overwhelmed by rocks and residing almost beneath the threshold of perception, unless one specifically focuses on it. This state of things in itself plays out a curious cognitive trick: for if life is all around but hardly to be seen, just how far—and in what "direction"—beyond the threshold of human perception does it extend? The land feels like it is pulsating with invisible life, so does it not make sense that it is there even if we cannot see or identify it? Where are the limits of the emotional sensing of life?

We did not spend our days on the tundra reflecting on such high-level ontological and epistemological questions, but they surfaced more or less clearly as we moved around the landscape. We had a laid-out plan for how to survey the key area, but in practice the terrain itself guided our movements to a substantial degree; we were not walking in lines but trying to find a manageable way of navigating this demanding terrain. While doing so, it was difficult to picture any extensive human engagement with Ritničohkka, at least in the summertime—though presumably the fjell is not an any less difficult or hostile environment in wintertime, in the purely practical terms of coping there. While we were initially worried about whether we would actually find anything of direct archaeological interest in this recently unfrozen sea of boulders, the discovery of reindeer bones, both where the snowfield had once been and outside that area, put that worry to rest.

One reason for the contradictory feeling of life and lifelessness was, as mentioned, the seeming absence of living animals. As the landscape is part of the Sámi reindeer herders' seasonal pastoral area, we had expected to encounter more life here. Then again, our attention turned mainly to our feet, as we took utmost care not to hurt ourselves, which made us highly focused on and attentive to the terrain. It is worth reiterating that, to move safely here, we really had to pay attention, as if learning how to walk for the first time. This attention got us deeper into the landscape, with its structures, features and components, and we started spotting things—including finds—that might otherwise have escaped our attention.

This is why, somewhat astonishingly, as we were carefully descending down the steep slope, Oula spotted what appeared to be a tiny stain of rotten wood by a boulder and, in association with it, what looked like a worked piece of quartzite. Quartzite does not occur naturally in this region, and on closer inspection it turned out to be our most straight-up conventional archaeological find, supplementing the reindeer and other bones that we collected (to be published in Fjellström et al., forthcoming).

But while we were on the fjell encountering reindeer bones that represented different anatomical parts of the animals, new questions arose. How exactly did the reindeer bones end up here? While there is nothing inherently strange about finding reindeer bones lying on the ground, it is also not transparently clear with what kind of past events and processes the bones have been entangled. However common and natural the reasons for the bones being up there may have been, experientially they contradicted the perceived lack of larger animals in the landscape. The bones scattered among the rocks somehow underlined the absence of perceivable animal life, the presence of death in the form of bones and the idea of invisible life imbued in this land. They hinted at a secret, hidden life unfolding in relation with this land, without our being able to perceive it. There was a resonance to the fossil-like formations we spotted in the bedrock, which suggested a very ancient and alien past that, our imagination told us, had fused into the bedrock, hence echoing how the reindeer bones had become part of this peculiar landscape.

4.5 Reindeer, Mosquitoes and the World on the Move

In recent times, Ritničohkka has been the setting, however peripheral, of human activity. This activity ranges from the construction and maintenance of the telecommunication antenna and the associated cabin to occasional visits by Finnish border guards, hikers and hardcore off-track skiers and snowboarders. But first and foremost, Ritničohkka has been part of the Sámi reindeer herding landscape for centuries, and it was likely a reindeer hunting landscape even before then. The (pre)histories of human interaction with reindeer in this environment, and in relation to the snowfield in particular, were the primary reason for our fieldwork on Ritničohkka.

While conducting the survey, as we started wondering about the stories of the scattered bones in the landscape, we began asking deeper questions. In general, the bones appeared to lack butchering marks or other signs of human manipulation, but there were also few clear signs of predator attacks. What we seemed to have, then, were scattered reindeer bones from different body parts lying here and there in the stone sea, and it was impossible to tell, on purely empirical grounds, how and why exactly the bones had ended up on the fjell. They could have been the bones of animals that died from natural causes and were then redistributed by various forces of nature. Bones can also indicate predator or human activity even if this often cannot be determined from the bones themselves. Some find locations were on very steep and rocky slopes where the reindeer, even though they can move about on difficult terrain, would conceivably have little reason to venture to. So we figured that postmortem events likely moved the bones to those places. Although there is nothing extraordinary about the bone finds we made, no deep mysteries about them, in the sense that reindeer bones are commonly found in reindeer herding landscapes, the bones nonetheless inspired speculations about their stories and histories (Fig. 4.5).

Whatever the specific stories of particular bones, the Ritničohkka snowfield likely played a role in them insofar as it has been a meaningful place for the reindeer. As such, it was also a meaningful place for reindeer herders. Perennial ice and snow patches on the mountains are important for the reindeer because, during the summer, in an open landscape largely devoid of shade, the cold radiating from the frost offers respite from the heat. The melting water supports the growth of vegetation,

4 MOUNTAIN BEINGS   93

Fig. 4.5 Fresh looking yet centuries-old reindeer bones recently uncovered from under the ice and snow at Ritničohkka

hence providing food for the reindeer. But most importantly, the reindeer are drawn to summertime snowfields and the open and windy high grounds of the fjells in general to escape the scourge of mosquitos and other insects. Insects, and mosquitoes especially, tend to prefer heat to cold, making them avoid cool air. As a result, mosquitos have had an important impact on the selection of seasonal pastures by the reindeer.

Furthermore, mosquitoes contribute to the herding behaviour of the reindeer, as herding also provides some protection from mosquitoes—which in turn cultivates behavioural traits that make the reindeer more herdable by humans (Helle 2015). Indeed, there is a Sámi story of the mosquito as an ambivalent (indeed monstrous) reindeer herder, recognizing the role of mosquitos in shaping reindeer behaviour but also the fact that they can exhaust the reindeer to the brink of death (Helle 2015). Although the prevalent local human strategy for coping with mosquitos is to pretend that they do not exist, it could be speculated that the suffering of the reindeer from these pests bridges the sympathy-gap between human and reindeer, since they both have to endure the same ordeal. And, as an old Sámi joik muses, it would be difficult to herd the reindeer at all in the summertime without the help of the tiny mosquitoes and other plaguing insects: "Mosquito drives the reindeer to the fjells, herds the reindeer together out from the valleys. You wouldn't get along with the reindeer if no mosquitoes existed. It is a small creature, but in front of it a large herd will escape" (Paulaharju 1922: 99, our translation).

For outsiders especially, mosquitos are a terrible nuisance, as made explicit in various historical accounts. This is reflected in the diaries of German soldiers based in Lapland during the Second World War (see Norum et al. 2021). In an earlier account, the Italian naturalist Giuseppe Acerbi, crossing the watershed of Northern Lapland near Kautokeino on foot, writes, in a typical manner featuring the common prejudice of European travellers towards the local Sámi,

> Though it was now drawing towards midnight, the torment we suffered from the musquetoes [*sic*], instead of being abated was increased. The night was perfectly calm, and the insects attracted by the effluvia of our Laplanders, pursued us in our course, surrounded us, and involved us as in a cloud. (Acerbi 1802: 50)

Though more than two centuries have passed since this acerbic blaming of local guides for the mosquito swarms in the far North, little has changed about some of the attitudes prevalent in Lapland tourism.

From a cultural perspective, mosquitos are malicious beings. Most people who engage closely with them probably have mainly negative opinions of them. It is easy to see how their pervasive presence might drive people—and reindeer—mad, so perhaps steadfastly ignoring them is a sensible way of coping with them. What is curious about mosquitoes in this context, however, is that they can be regarded as monstrous beings in several respects. Most obviously, mosquitos suck blood, and blood is the very real essence of life and death. Their presence in the landscape is also peculiar in that they seem to be omnipresent during the mosquito season, hovering like little damned, monstrous fairies, making a most irritating sound, which indeed is one of their main characteristics. Although mosquitos can be seen, they are often sensed as an unlocatable sound, which is particularly unnerving in a dark cabin in the middle of the night, as even a single mosquito can effectively ruin a night's sleep.

Moreover, mosquitos are curiously invisible in Finnish folklore even though they have been a human companion in the North for a long time, which raises questions about how mosquitos have been perceived and understood in the past. Culturally speaking, what kind of beings are and have mosquitoes been? For instance, to what degree does it make sense from a human or reindeer point of view to regard mosquitos as individual beings? Should they perhaps instead be considered primarily as swarms? A swarm of mosquitos is certainly an actual thing, an entity, but it also further highlights the monstrous characteristic of mosquitos, as the presence and boundaries of such an entity are rather more elusive and indefinite than that of a single mosquito. Also, both individual mosquitos and a swarm of mosquitos can have quite an ethereal presence—visible, invisible and transparent simultaneously, depending on the specific conditions. Yet, at the same time, mosquitos are almost like a connective tissue, something that sheds light on the entanglements of particular places in the landscape, people, animals and plants.

Like the reindeer who escape from mosquitos by climbing up the fjells, fieldworkers also largely try to avoid the scourge of mosquitos. Due to the favourable season, we encountered few mosquitos on Ritničohkka. But we also did not see a single reindeer. The absence of these two iconic figures of Lapland also contributed, however subtly, to the sense of the simultaneous presence and absence of life and death in the landscape. Yet, as the bone finds indicated, reindeer had been present, drawn by the now-disappeared snowfield. It is, of course, impossible to tell yet if, or to what degree, the disappearance of the snowfield has affected and will affect the reindeer's relationship with the place and landscape. We do not know what specific ecological or other impacts the disappearance of the snowfield will have, but it has potentially altered and restructured fundamental relationships between Ritničohkka, the reindeer and the local people. The disappearance of the perennial ice may even make the fjell a dead zone in the reindeer landscape and mindscape—a ruin deserted by the reindeer, and therefore by reindeer herders as well. If so, the reindeer bones that we collected from the fjell have an added eerie symbolic meaning to them.

The eeriness that we grappled with was only exacerbated by the condition of the shelter on our increasingly desolate summit.

References

Acerbi, J. 1802. *Travels Through Sweden, Finland, and Lapland, to the North Cape, in the Years 1798 and 1799*, vol. II. London: Printed for Joseph Mawman.
Barad, K. 2017. No Small Matter: Mushroom Clouds, Ecologies of Nothingness, and Strange Topologies of Spacetimemattering. In *Arts of Living on a Damaged Planet: Ghosts and Monsters of the Anthropocene*, ed. A. Tsing, H. Swanson, E. Gan, and N. Bubandt, 103–121. Minneapolis: University of Minnesota Press.
Barry, F. 2007. Walking on Water: Cosmic Floors in Antiquity and the Middle Ages. *The Art Bulletin* 89 (4): 627–656.
Bernbaum, E. 2022. *Sacred Mountains of the World*, 2nd ed. Cambridge: Cambridge University Press.
Breda, N. 2022. Are Stones Living? Lagoonscapes. *The Venice Journal of Environmental Humanities* 2 (1): 163–176. https://doi.org/10.30687/LGSP/2785-2709/2022/01/009
Cunliffe, B. 2017. *On the Ocean: The Mediterranean and the Atlantic from Prehistory to AD 1500*. Oxford: Oxford University Press.
Davidson, P. 2005. *The Idea of North*. London: Reaktion Books.
Haanpää, P. 1980. *Teokset 5*. Helsinki: Otava.
Harman, G. 2012. *Weird Realism: Lovecraft and Philosophy*. Winchester: Zero Books.
Hazen, R.M., D. Papineau, W. Bleeker, R.T. Downs, J.M. Ferry, T.J. McCoy, D.A. Sverjensky, and H. Yang. 2009. Mineral Evolution. *American Mineralogist* 93: 1693–1720.
Helle, T. 2015. *Porovuosi: tutkija pororenkinä Sompiossa*. Helsinki: Maahenki.
Herva, V.-P. 2021. Minoan Lapland: Fieldwork, Spirituality and Connecting Across Time and Space. *Time and Mind* 14 (3): 397–415.
Herva, V.-P., A. Varnajot, and A. Pashkevich. 2020. Bad Santa: Cultural Heritage, Mystification of the Arctic, and Tourism as an Extractive Industry. *The Polar Journal* 10 (2): 375–396.
Herva, V.-P., O. Seitsonen, I. Banks, G. Moshenska, and T. Paphitis. 2024. Folk Magic and the Haunting of the Second World War in Finnish Lapland. *Cambridge Archaeological Journal*: 1–19 (online first). https://doi.org/10.1017/S0959774323000495.
Ingold, T. 2000. *The Perception of the Environment: Essays on Livelihood, Dwelling and Skill*. London: Routledge.
Ingold, T. 2019. The North Is Everywhere. In *Knowing from the Indigenous North: Sámi Approaches to History, Politics and Belonging*, ed. T.H. Eriksen, S. Valkonen, and J. Valkonen, 108–119. Abindgon: Routledge.
Naum, M. 2016. Between Utopia and Dystopia: Colonial Ambivalence and Early Modern Perception of Sápmi. *Itinerario* 40 (3): 489–521.

Norum, R., V.-P. Herva, and M.O. Lundemo. 2021. Encountering/Thinking Mosquitos. *Time and Mind* 14 (3): 417–430.
Olsen, B., and Þ Pétursdóttir. 2021. Writing Things After Discourse. In *After Discourse: Things, Archaeology and Heritage in the 21st Century*, ed. B. Olsen, M. Burström, C. DeSilvey, and Þ Péturdsóttir, 23–41. Abingdon: Routledge.
Paulaharju, S. 1922. *Lapin muisteluksia*. Helsinki: Kirja.
Paulaharju, S. 1927. *Taka-Lappia*. Helsinki: Kirja.
Petersen, D.O.J. 2017. At the Mountains of Monstrosity: Towards Ecomonstrous Entanglements Through Images of a Fjord. *Women, Gender & Research* 2–3: 70–88.
Piatti-Farnell, L. 2021. Arctic Gothic: Genre, Folklore, and the Cinematic Horror Landscape of Dead Snow (2009). *Studies in European Cinema* 18 (1): 76–85.
Valtonen, T. 2019. Miten Saanasta tuli pyhä? Erilaisten rinnakkaisten Saana-diskurssien tarkastelua. *Terra* 131 (4): 209–222.
Westerdahl, C. 2005. Seal on Land, Elk at Sea: Notes on and Applications of the Ritual Landscape at the Seaboard. *International Journal of Nautical Archaeology* 34 (1): 2–23.

Open Access This chapter is licensed under the terms of the Creative Commons Attribution 4.0 International License (http://creativecommons.org/licenses/by/4.0/), which permits use, sharing, adaptation, distribution and reproduction in any medium or format, as long as you give appropriate credit to the original author(s) and the source, provide a link to the Creative Commons license and indicate if changes were made.

The images or other third party material in this chapter are included in the chapter's Creative Commons license, unless indicated otherwise in a credit line to the material. If material is not included in the chapter's Creative Commons license and your intended use is not permitted by statutory regulation or exceeds the permitted use, you will need to obtain permission directly from the copyright holder.

CHAPTER 5

At the Basecamp

A cabin
at the edge of the world
Where nothing
moves in human time
except the creeping mould
And the emergency rations
turn black
 —"*Poe*tic Pastiche" #2, Aki Hakonen

5.1 First Encounter

The cabin near which we initially landed on Ritničohkka and which we had decided to make our base came to mediate our perceptions of and relations with Ritničohkka in important ways. There were two cabins on top of Ritničohkka: the first a living quarters and the second a large shed, housing firewood and assorted chemical containers and machinery. Once we had landed and made it over to the first cabin, what transpired was the first unanticipated event of our journey. The key, provided to us by border guards, did not work. Even with everyone taking turns to try this and that, the key would not turn in the lock, no matter what. The cabin that was supposed to be our shelter was unwilling to do our bidding. We figured that breaking in should be easy. But for a remote cabin scarcely in use and with nothing of any value inside, its door was uncannily heavily bolted shut. In hindsight—if only jokingly or metaphorically—the tight sealing of the cabin and our difficulties in entering it could have been taken as a portent or warning (Figs. 5.1 and 5.2).

Such portents and premonitions feature amply in northern folklore, even today, and, rather than idle "superstitions", are better treated as reflecting a deep awareness of the environment and how its different constituents are interrelated (Herva and Seitsonen 2020; see also Herva and Ylimaunu 2009; Herva 2010). We had been informed that the cabin had originally acted as a well-furnished and comfortable service and maintenance cabin for a telephone and radio tower in the 1980s and 1990s, with electricity and gas heating, but that the tower had not been in active use since then and had been turned into a link in the official government emergency radio network. This likely explained the heavy and unbudging lock on its door (Fig. 5.3). The tower is still maintained, though it was not operational when we arrived, as indicated by the severed cables that we observed next to it. After the tower stopped being in active use, the cabin was left unmaintained, and a slow ruination process took over,

Fig. 5.1 The cabins at sunset on the first evening

Fig. 5.2 Roger attempts to open the lock while others carry bags and bundles of gear from the landing zone. On the right, one of the cairns built next to the cabin

with only Border Guards now and then visiting the place during their wintertime patrols.

The door of the shed opened without any problem, but this structure could not be heated and was otherwise unsuitable as a comfortable overnight shelter. After some failed attempts, we eventually got into the cabin through a window that we managed to open without breaking. It was early afternoon and the weather was beautiful with sunshine, a light wind and temperatures around a comfortable 10 degrees Celsius, so our difficulties with the cabin were mostly amusing at this stage. However, as we settled into the cabin and familiarized ourselves with it, we no longer felt quite so amused.

We knew that the cabin had not been used for a long time and were in principle prepared to camp in tents if the place was uninhabitable. Our first impression was that the cabin was indeed in disorder, but it did not initially seem unmanageable. Once upon a time, the cabin had been remarkably well equipped for such a remote place, with electric lights, gas heaters, cooking facilities and a good stove, as well as adequate bedding and mattresses to facilitate the comfortable maintenance of the radio and telephone antenna located on the top. However, those days

Fig. 5.3 There may have been a reason why the door to the cabin was bolted tightly shut

were long gone, and the last time the cabin had been properly used was nearly two decades ago. The closer we engaged with the cabin, the more disturbing it turned out to be. Sure, there was general disarray, which was unsurprising. But the place—the walls, the floor, the mattresses on the beds—turned out to be rotten and mouldy, dark and oppressing. The temperature inside was several degrees colder than outside. Also, the stove was rusted through on all sides and had become almost unusable, which effectively made properly heating the cabin impossible.

We did not think about it right then and there, but the cabin, with growing mould and its wooden structures decaying and rotting in place, surrounded us with weird and threatening, practically invisible, lifeforms, as in the case of fungi, which can be considered emblematic beings when it comes to "weirding", as well as central to the imaginaries of the Anthropocene (Tsing 2015; Bradić 2019). A sealed plastic box labelled "Emergency Rations" was filled with black fungus growing inside, with all the inner walls moist and covered by mould (Fig. 5.4). The lid of the box revealed that the rations had been taken to the cabin in 1979, and stated that "if you need to use anything, inform the Muonio tele office what you have eaten", referring to an office that itself has ceased to exist several decades ago; later, someone had added a sarcastic note on the lid: "I wouldn't open this". The mattresses on the bunk beds in the cabin were not only mouldy but also thoroughly soggy. On taking shelter in the cabin, we were to co-inhabit it with many unwanted and disturbing non-human lifeforms. Indeed, the cabin itself became the first monster we encountered. And we were stuck with it.

We tried our best to clean the cabin, hauling out the disgusting pieces of old bedding and the rotting foodstuff, and storing the worst of the decomposed furniture in the shed (Fig. 5.5). Opening all the ventilation and keeping both the window and door open to air the place out provided some relief. As the evening turned to night, we were hanging out outside the cabin, roasting sausages for dinner. The wind was picking up, and we knew that the weather was about to change. Stormier weather was forecast, but it was impossible to know what form it would take, which caused a slightly nervous anticipation of what was to come. Ritničohkka

Fig. 5.4 The emergency rations in the cabin turned out to be several decades past their expiration date

Fig. 5.5 Clothes hanging inside the cabin. Electric cables and bunk beds tell stories of the cabin's glory days; the thoroughly mouldy mattresses were thrown into the storage shed

has two summits 750 metres apart, and the cabin was located on the eastern summit. The summits are practically on the same level, but our basecamp was decorated with a high-rising metal antenna, which would undoubtedly direct any lightning uncomfortably close, as it stood only a few metres from the cabin.

5.2 A Mobile Dead Zone: Roger's Account

When the helicopter dropped us off at the top of a fjell in what felt like the middle of nowhere, we stepped out, lugged our bags away from the chopper and watched the pilot take off into the gentle breeze. We were alone. I had been filming a video with my smartphone during the 15-minute ride from Kilpisjärvi, and my phone was warm. I looked down at the screen. No bars. Searching. No bars. Still searching. I thought to myself that the lack of signal was due to interference from the helicopter radio, even though by this point the pilot was well into his trip back to Kilpisjärvi. A hike up to the cabin, and there too: nothing. I walked around. Nope. Dead. A dead zone.

"How could this be?" I wondered. This is Finland. In fact, this is the very top of Finland, the country's highest point. Better: this is the border between Finland and Norway, two of the most technologically developed countries on the planet. There were no obstructions anywhere. A clear line of sight between us and the atmosphere all around us. How was it possible that there was no mobile signal here? The experience of travelling in other parts of the world suggested that, surely, there would be a mobile signal up here, if anywhere.

Dead zones are locations where there is limited or no mobile technology access or data signal coverage. The phenomenon of spaces of virtual emptiness or lack is not uncommon in many rural areas, such as protected nature reserves, stretches of wilderness or even farmlands located far from mobile data (3G/4G/5G) antennas or where the view of satellites is obscured that would otherwise provide GPS services to mobile devices on the ground.

The ubiquity of internet and data connectivity has defined contemporary social life in many parts of the world today. After the early years of mobile internet in the mid-twenty-noughts, a generalized expectation swiftly took over that suggested that anyone who possesses a laptop or phone also possesses the right to ever-present web access, messaging services, GPS and telephone connectivity. One of the noughties' most well-known television advertisements was for the US mobile provider Verizon Wireless. It ran from 2002 to 2011 and featured a middle-aged man, played by actor Paul Marcarelli, who moves around to various places in pursuit of a data connection, calling out loudly into his phone, "Can you hear me now?" The refrain became a catchphrase for that entire decade and the one that followed, and it was indeed already one that many mobile phone users could relate to, given that at that point in time, when the development of mobile connectivity infrastructures was still largely in its infancy, it was not uncommon to enter spaces with immediate or gradual data drops, in rural as well as urban spaces.

Such dead zones still exist in many parts of the world—particularly in developing countries and in areas far from urban centres that have yet to receive adequate technological data infrastructure. Mobile technology plays an ever-increasing role in the experience of mobile people, from tourists (Brown and Chalmers 2003) to migrants (Zijlstra and Van Liempt 2017), though there tends to be an implicit assumption that tourists embrace mobile connectivity. There persists some ambiguity about the use of mobile technology in tourism, and in the value of connectivity versus the desire to "get away from it all" (Dickinson et al. 2016). Nevertheless, finding oneself in a dead zone can be a stressor related to social communication, the sense of safety versus the longing to escape, and to being able to be fully engaged with one's present company and setting (Lang and Borrle 2021; Pearce and Gretzel 2012).

As Adrian Tanti and Dimitrios Buhalis have eloquently outlined:

> Try and visualise a traveller who is about to embark on a journey. As he makes his way to the airport, he listens to music on *Spotify*, keeps socially updated on *Facebook*, and completes the online check-in for the flight. He boards the flight with a boarding pass that was retrieved on a smartwatch, arrives at the destination, and books a ride with *Uber* to the city centre. After checking-in at the hotel, the visitor selects and books a restaurant through *TripAdvisor*, and then navigates to it using *Google Maps*. Once the food arrives, he captures a photo and uploads it on *Instagram* and *Facebook*, shares his location and writes a short insight on *Twitter*, chats on *WhatsApp* and reviews the restaurant on *TripAdvisor*. *Google Now* suggests a list of attractions in the vicinity that he might want to visit. Once in the recommended attraction, he opens *Periscope* and shares a live stream of the view with people from all around the world. Current technology allow for all these activities to be performed effectively and efficiently. (Tanti and Buhalis 2017)

Though slightly dated (and gendered), the point the authors make in this vignette is clear: technology is an integral aspect of the experience and imagination of contemporary travel.

It is a tad perverse to consider that the reason nature tourism in rural and remote parts of the world is so popular is that it lures us with the possibility of "getting away from it all", even as an expectation of connectivity persists among many such visitors. This is arguably just one sign of our selfish, on-demand society, one in which we would like to be able to have what we want when we want it; and, even when we say that we don't want it, we would kindly like to reserve the right to instantaneously demand it. Searching for bars, I realized how attuned I was to this expectation of connectivity.

Approaching the cabin from a distance, there is only one landmark that demarcates its location through the clouds: the telecommunications antenna, which sits on the cabin's west side. That such a defining semaphore of modern life and convenience could be located in a mobile dead zone feels blatantly contradictory. While the large antenna that currently resides at the cabin facility used to be owned and operated by the Finnish telecommunications company Telia, it is no longer employed to transmit civilian, consumer mobile signals—hence the dead zone in and around the cabin. The antenna is used by the public safety network Virve, which enables Finnish authorities and other mission critical operators (e.g. the Finnish military, police and emergency response centres) to communicate effectively and safely. This explains why, despite the cabin being next to a large telecoms antenna some 15 metres in height, there was no mobile signal to speak of in the immediate vicinity.

5.3 Spiritual Communications

Pondering on the weather and struggling to keep a fire on the exposed spot in strengthening wind, we constructed an ad hoc sheltered fireplace that grew more substantial and higher as the evening unfolded and more stones were added to the construction. There were two large stone cairns that somebody had carefully built next to the cabin, a practice which seems to have become more common on the fjells over the last decades, as tourists build cairns to mark their presence. However, cairn building also has much more ancient roots and slightly unnerving associations and resonances. We could not assess their age or purpose, but the deposition of a few antlers and the remnants of metal objects in the cairns from the construction activities surrounding the cabin was reminiscent of the practice of making offerings at Sámi sieidi offering sites (Äikäs 2015).

It has become more and more common for hikers to construct small towers or piles of rocks on fjells when they summit them. Usually a hiker climbing a summit contributes only a single rock to the monument, an act seen as a form of communion with others who have reached the summit in the past and with those yet to arrive. But the cairns by the Ritničohkka cabin were much more substantial than the makeshift piles that hikers ordinarily make. The objects between the rocks of the Ritničohkka cairns—whatever the reason these particular cairns had been built and the objects placed in them—readily brought about an association with northern spiritual practices originating in prehistory. There is, on the one hand, the Sámi tradition of making sacrifices at particular places in the landscape, known as sieidi. On the other hand, cairns have been constructed for various purposes since the Neolithic and especially the Bronze Age, and have long been associated with the supernatural, irrespective of their age and original function. One aspect of both sieidi and cairns is that they are about communication with otherworldly realms beyond the here and now. On Ritničohkka there is thus a weird connection between prehistory and modern technology, materially and monumentally manifested by the proximity of the cairns to the telecommunications antenna.

Sieidi (NS; Fi. *seita*) are sacred Sámi sacrificial sites where diverse offerings have been made from at least the sixth–seventh century AD onwards (Salmi et al. 2015). Some sieidi are still used today as places for making offerings by locals, tourists and people subscribing to neo-pagan and neo-religious ideas and beliefs (Fonneland and Äikäs 2023). Sieidi take various material forms and are typically connected to specific natural sites, such as peculiar rocks, trees, waterfalls or topographic formations like mountains and cliff faces. Activities at sieidi sites mirror the traditional Sámi worldview and the reciprocal relationships between human and non-human actors. Traditionally, people interacted at different places with sieidi and made offerings for numerous purposes, for instance, to maintain good relations with the land, animals and spectral entities, and to seek protection, blessing or luck in various activities such as fishing, hunting or herding. A wide variety of things have been offered at sieidi sites through time, including animals and animal parts (e.g. reindeer antlers, fish and the meat, bones and blood of other animals). Other acceptable sacrifices include wooden, bone and metal artefacts (the latter especially as far back as during the Viking Age) and also, later on and up to the present, tobacco and alcohol (e.g. Äikäs 2015; Salmi and Seitsonen 2022).

Sieidi use and other traditional Sámi religious and spiritual practices were suppressed when Christianity spread into the region, especially from the seventeenth–eighteenth century AD onwards. This can be seen archaeologically, for example, in the waning number of radiocarbon-dated bone finds from sieidis after the 1650s AD (Salmi and Seitsonen 2022). However, the use of sieidi continued, as indicated by both transgenerational oral histories and finds from more recent times, including coins and other artefacts dating all the way to the twenty-first century. This illustrates the resilience of the Sámi cosmology and worldview despite colonization, assimilation, modernization and other external pressures brought along with the Nordic nation states. The endurance of traditional practices and beliefs shows the vitality of Sámi values and culture and its deep connections to their own lands and environment (e.g. Fonneland and Äikäs 2023).

Ritničohkka is not—or at least is not publicly known to be—a sieidi site, and the cairns by the cabin are likely related to something other than spiritual practices. They might have been constructed for the purposes of map-making or as border markers, but they may also have been built by hikers, although that seems unlikely given the remoteness and markedly built environment of the summit. The tourist practice of building cairns on top of fjells tends to annoy locals because these piles have colonialist connotations as the marks of outsiders "conquering" the landscape. The practice also bothers some archaeologists, who have observed that the contemporary practice can result in the destruction of old cairns or make identifying them in the landscape difficult. There are cases where irritated hiking activists have destroyed supposed "tourist cairns"— as hardcore hikers want to be associated with casual tourists about as much as scientists do—which later turned out to be centuries-old trail markers. Ironically, the notion of the "untouched" northern wilderness is a distinctly colonialist view of Lapland.

Closer to Kilpisjärvi village, for instance, there is a historical chain of trail-marker cairns associated with an ancient trade route that ran from the northernmost Baltic Sea to the Arctic Ocean on the Norwegian coast. Another chain of cairns some 60–70 kilometres south of Ritničohkka marks the ancestral boundary, based on the unwritten common law, of the local Sámi *siidas*' land use catchments (Korpijaakko 1989). Even if, prior to the emergence of colonial outposts, the lands were not owned in a Western sense, there were traditional limits for the uses of the landscape based on the common law. As one of our Sámi interviewees noted: "In the old times, every patch of land had been commonly agreed upon [based on the common law], who can use it, for what and in what kind of ways".

But, as mentioned above, there are also metaphysically unsettling and "darker" associations to cairns, as Finnish folklore makes abundantly clear. This is, first and foremost, due to the association of cairns with the dead. Indeed, many prehistoric Bronze and Iron Age cairns have proven to be burials and drowned people were occasionally buried in coastal cairns as recently as the nineteenth century (see further Tuovinen 2002; Okkonen 2003; Westerdahl 2002, 2005). Cairns have also been loci of sacrificial practices, from prehistoric to historical times (e.g. Muhonen 2008). Knowing about these traditions of cairns as communication devices or channels with the dead and underworld—themselves associated with the High North in Finnish and Sámi traditions—made the cairns loom quite eerie when the top of the fjell became covered in clouds during our first night there, which is how it would remain for the rest of our stay. The idea in Finnish folklore that cairns are in some sense portals to the underworld is interesting here also due to its shamanistic allusions. Shamans venture into the underworld figuratively and metaphysically, but also literally, since cave-like formations have been conceived as places for accessing otherworlds and special—arguably shamanistic—knowledge (see e.g. Lahelma 2007; Vesajoki 2021; Rainio and Hytönen-Ng 2023).

Symbolically, at least, the two cairns struck some resonance with our attempts to engage—in a sense "communicate"—with the corpse of the snowfield and deceased reindeer and with a gone but spectrally present past. Small wonder, indeed, that archaeology as a practice is prominently associated with the supernatural and otherworldly forces in popular culture—in the Indiana Jones movie saga, for instance. While, substantially, we cannot say anything specific about the age or origins of the cairns or the objects found in them, the "offerings"—pieces of reindeer antlers and bones and bits of wood and metal—are curiously similar to "actual" offerings at traditional sieidi sites, and at least superficially can be taken to reflect nonmodern northern mindsets, cosmologies and mythologies and their entanglements with the landscape. But there are also related entanglements with the spiritual and the weird in regard to modern technology, as we will discuss next.

5.4 Uncanny Engagements with Technology

As we had only three full and two partial days to spend on Ritničohkka, we started our survey as soon as we had managed to enter the cabin and had cleaned up the worst of the mess that we encountered inside. After exploring the remains of the snowfield on the fjell's north-eastern slope, Markus and Aki set up their tents near the cabin, where the boulder field allowed it, while the rest of the crew decided to stay in the uninviting cabin. The sleep in the cabin was restless. It was cold, the cabin smelled foul, the strong wind whispered and occasionally howled in the chimney, and the cabin creaked, shooshed and rustled through the night. The cabin felt almost as if it were breathing, slowly and deeply. The cabin's antics roused a horror-esque feeling. Vesa-Pekka, Oula and Roger lay next to each other in sleeping bags on the dirty, mouldy floor covered with an aluminium space blanket, listening to the night's sounds, until each of them finally fell asleep.

During the night, when Oula heard nature's call and went outside, the wind had picked up but the scenery was still mostly clear. The blood-red sun hung low above the horizon beneath a thin veil of clouds racing across the sky (see Fig. 5.6). This ominous scene was an apt prelude to the weather that followed.

For, in the morning, a storm was upon us, and the top of the fjell was completely shrouded in a thick grey cloud. All colour had faded into shades of grey, and the visibility was down to less than ten metres. Our surroundings merged with the clouds only a short distance from us in every direction. The world had shrunk down to a hazy bubble, with the cabin at its centre. The wind had picked up during the night and grown very strong by morning, and yet, eerily, nothing besides the air seemed to move in that grey world. The cloud that enshrouded us appeared stationary, like it existed in some other dimension, undisturbed by the wind. Walking on the rocks, we could definitely feel the tackling wind attempting to get a hold of us. Though it was not exactly raining, the air was so full of moisture that it seeped fast into the fabric of all unprotected clothes.

Wandering off into the cloud for a survey in the buffeting wind did not make much sense, so we decided to wait for a couple of hours to

Fig. 5.6 Blood-red midnight sun hanging low behind a wisp of clouds. Silhouettes of the cairns in the foreground

see if the weather would change. Roger had downloaded a few films on an iPad and we agreed to waste battery time while trying to heat the cabin, the latter turning out to be in vain. The setup—the five of us sitting around a small table wearing padded outdoor clothes to fight the cold and watching some American film on a tablet in a rotten cabin on the summit of a remote fjell while a heavy storm howled around us—felt exceedingly unreal. Aki spontaneously called it "just wrong"—and "wrongness" is indeed a defining characteristic of the weird for Fisher (2016) (Fig. 5.7).

We had brought a few stacks of firewood with us, and there was also plenty of soggy firewood in the shed. The fireplace had rusted through and partly collapsed, but it was possible to build a small fire there once we lined it with some rusted pieces of sheet metal foraged from the shed. We reckoned that using the fireplace was fine as long as it was not left unsupervised. However, keeping the fire didn't do much, as the fireplace turned out to behave weirdly, failing to do what fireplaces are expected to do: provide warmth. No matter how much wood we burned, the temperature in the cabin never went above 11 degrees—it was as though the fireplace, too, was dead deep inside, despite the living fire dancing in it.

Fig. 5.7 Movie matinée in the cold cabin while waiting for the bad weather to pass

The cognitive dissonance of watching an action film—a poor one, it turned out—on a tablet in the circumstances in which we found ourselves was near nauseating. Simultaneously, however, it resonated with our general feeling of things working—or not working—weirdly or very erratically. This, in turn, struck another chord with northern worlds as lived environments, with messy more-than-human relations being integral to northern ways of life (Herva and Lahelma 2020). This took various forms or expressions at the cabin. For instance, we discovered that there was indeed random mobile data connectivity which generally did not work inside the cabin, but there was a fairly stable hotspot, limited to a very specific area, some twenty metres from the cabin, out in the elements. This hotspot could be imagined, especially in the evocatively grey and stormy conditions, as something of a vortex accessible only at very specific space–time coordinates, much like the one seen in *Twin Peaks: The Return* (2017). Although invisible, it was easy to envision the hotspot as a portal or a wormhole opening into the cloud that enshrouded us and connecting our weird world on the fjell to our accustomed world of Wi-Fi and smartphones.

More generally, the cabin, the shed and the telecommunications antenna felt like they were out of place in our odd pocket of reality, but also tangible meeting points between two metaphysically and experientially different worlds. There was the world we originally came from and there was this other world vibrating with monstrous dissonance and otherness, but in a manner that lingered somewhere beneath the surface of the landscape and was anchored in certain peculiar features, such as the curious rock formations. Indeed, the place felt very much like a cabin at the end of the world (Sect. 5.6) or like the mysterious abandoned factory in the midst of a shapeless desert in Thomas Ligotti's short story *The Red Tower* (1996) (Fig. 5.8).

The telecommunications antenna on Ritničohkka is itself a clear and monumental symbol of high modernity and connectedness, and of how technological modernity has penetrated geographically remote environments. Yet, there was a curious dissonance at play here, as Roger already mused before, for, despite the prominent physical presence of the antenna, the lack of internet connection was palpable. Like the fireplace in the cabin, the antenna on Ritničohkka was, in fact, a thing that was not properly "alive", though not completely dead either, but more like a kind of a zombie, metaphorically echoing the cairns next to it as latent connectors to otherworldly realms. As explained earlier, currently the antenna

Fig. 5.8 Stone cairns in the mist on the fjell's high point, adjacent to the deceitfully inhabitable-looking shed

is not part of public consumer telecommunication systems but of the state's emergency communications network. It was connected by cables to various electronic devices inside the storage cabin, which we decided not to touch. The antenna, with the associated devices, provoked a slight sense of mystery in the overall setting of Ritničohkka, which brought to mind the recurrent theme in *Twin Peaks: The Return* (also featured in the 1992 Twin Peaks movie *Fire Walk with Me*) of electricity having supernatural properties that connect or function as a conduit between different dimensions of reality.

Indeed, modern information and communication technologies have diverse weird dimensions to them not only in fiction but also in the real world, which explains how and why computer programming and other engagements with high technology can be entangled with spiritual ideas and practices. Digital technology can work in uncanny ways and generate a sense of wonder, enchantment and connectedness to higher dimensions of being (e.g. Aupers 2009), and it can break down mechanistic causation and linear space–time. There are also other weird and monstrous aspects to digital technology, as exemplified by artificial intelligence and algorithms (regarding "algorithmic pollution", see Schultze et al. 2018).

5.5 Faecal Action

While some of us were more addicted than others to smartphones and constant online presence, Ritničohkka also presented us with some very primaeval and rudimentary basic questions of being in the world and place-making practices. We had, of course, brought all our food and an initial supply of water with us, but defecating needed to be arranged locally on site, which provided some unexpected insight into the ways we related to the local landscape and into the group's dynamics as well. A discussion of defecation practices might at first seem like a juvenile joke, but as demonstrated by Rachel Vanessa Lea's (2001) doctoral thesis, titled *The performance of control and the control of performance: Towards the social anthropology of defecation*, defecation practices are of paramount significance, though long overlooked as a subject of academic research.

Admittedly, our little excursion into the world of faeces on Ritničohkka started as a joke—and humour, as Lea (2001) points out, is indeed an integral cultural element of defecation. Besides the highest peak in Finland, Ritničohkka also boasts the highest outhouse in the country, located in the boulder field near the cabin. Only, this plank-built outhouse was effectively a ruin at the time of our stay, barely holding itself together in the tempestuous wind. Its door was off its hinge and the back wall had collapsed, and as the doorway faced south, the user of the outhouse would have been completely exposed to the driving rain and wind. We opted not to test whether the outhouse could still serve its intended function—not so much out of fear that the structure might collapse, but more out of disgust over the toilet's state of disrepair; Oula decided to investigate more closely whether the receptacle of the outhouse had been emptied—it had not.

But we still had to perform our routine bodily functions, which we quite likely would not have even thought about if the storm that began during our first night on the fjell hadn't imposed some significant practical limitations—at least as regards comfort and stability—to defecation. In many cultures, including Western society and no less Finnish society, the matter is highly regulated. Strict zones—private bathrooms, public toilets and the like—are set aside for the purpose. Even in most national parks and maintained hiking trails, outhouses comprise treasured basic infrastructure. Meanwhile, however, on Ritničohkka, each team member chose to improvise in his own way, which presented some interesting dynamics that we all recognized and discussed on site.

For people raised and attuned to perform the deed in closed confines, defecating in an open landscape is unusual. In our case, being inside a cloud with a visibility of less than 20 metres, it was deeply unnerving to wander far off. In navigating to perform the deed, the vicinity of the summit was key, as were a few unusually large boulders towards which everyone gravitated. Movement remained difficult, and the rocks were increasingly wet and slippery, with the unpredictable gusts of wind demanding attention.

After this unfamiliar transitioning to an accustomed practice, the main difference that emerged between us team members was a dichotomy between what might be called the "manifest" and the "hidden" approach to defecation.

A boulder field is made up not only of rocks and lichen but also of crevices and cavities—of empty space. Some of these cavities extend underground into darkness, with the distance to the bottom unknowable without a measuring tool. Such a portal offers an appropriate place to get rid of unwanted material. In fact, with some finesse, a unit of excrement can be deftly let go into the void, leaving no earthly trace of the activity. Toilet paper, an essential accessory from the nineteenth century onwards, can also be disposed of in such cavities with ease, leaving the performing agent free to continue on in quiet shame.

However, some team members, especially those with the highest academic standing, chose the opposite approach. These members tended to choose the most apparent landmarks, especially large boulders and marked their targets on open surfaces. This manifest approach, however, was not simply self-serving but also practical in the moment, since an appropriately sized and shaped boulder provided support in the slippery and windy conditions. Yet, somewhat graphic descriptions of the results—with expressions such as "sprays of shit" that "exploded" on vertical rock surfaces, "slithering down along the side following the contours of lichen" to "yo-yo on an overhang"—were matters of great pride and personal performative and informative joy when shared with the others, which readily shows that more than purely practical considerations operated behind the "manifest" approach. Indeed, the performative act may be understood as an exercise in power and prestige, belittling one's prestigious standing, since one can afford it. The principle is much the same as the basic potlatch model of destroying one's property to indicate that one can afford it.

The element of shame related to defecation seems ubiquitous, yet we learn it growing up. Humans are not born with such shame, nor, presumably, are other animals, but instead it is a cultural product. Mary Douglas, while mentioning defecation and faeces only nine times in her book *Purity and Danger* (1966), discussed the act of public defecation in the context of Hindu class society, where public defecation was such a taboo for higher classes that the existence of the activity was practically denied, even when seen performed in plain sight by the lower classes. In her seminal doctoral thesis, Lea highlighted the paradox that, in the Western world, defecation is regarded as a hidden and shameful practice while simultaneously it is a source of great public humour, which has contributed to defecation being a difficult subject to approach academically (Lea 2001).

For archaeology, of course, faeces are a matter of great interest because many issues relating to defecation are present at archaeological sites. While actual coprolites (that is, fossilized turds of excrement) stir excitement among archaeologists as potentially important sources of information about past lifeways, much less has been written about the organization and infrastructure of defecation, how these were managed and the cultural elements relating to the activity in inhabited places. For instance, latrines and their deposits can contain and offer loads of information about past lifeways, nutrition and disease, but how these facilities were spatially and mentally oriented, used and maintained has been less discussed.

One interesting aspect about latrines in Lapland that we recently observed is that the latrines used by German soldiers at the German-run Second World War prisoner-of-war camps were usually placed at the outermost edges of the camp, towards the (for them alien and threatening) wilderness but, importantly, also adjacent to the prisoners' accommodation areas (Seitsonen and Herva 2011; Seitsonen et al. 2017). This might mirror the conscious or unconscious effects of years of propaganda, hate-mongering and state-sanctioned policy in which intolerance and categorical discrimination were presented as scientific standards. In the Germans' minds, this propaganda might have likened the prisoners—mostly Soviet Red Army conscripts—to disposable rubbish or even to noxious toilet waste (Seitsonen 2021: 114). This serves as a warning in today's political atmosphere—with far right-wing values on the rise in Finland and elsewhere and increasingly insulting rhetoric being used against minority groups—that such associations, constructed verbally, matter and can eventually metastasize into a dark reality.

As for the act of defecation in the open wilderness, there is nothing inherently shameful or disrespectful about it. Faeces provide sustenance for flora, fauna and fungi, and can be understood as a contribution to the biosphere. Spiritually, the act may be conceptualized as an offering, whether hidden or displayed. The local Sámi often remark half-humorously that when you go to defecate outside a seasonal camp on the tundra and pick up a rock to cover your faeces, only to discover a fresh turd under it, it's a sign that it's time to relocate the camp.

5.6 The Cabin at the End of the (Other)World

It was our second—or first full—day on Ritničohkka. The stormy weather seemed to calm down somewhat by noon, and the few glimpses that we gained of our surroundings as the clouds shifted in the sky suggested that only the top of the fjell remained in the clouds. So we decided to continue our survey. We walked through the cloud, descended below it and spent the rest of the day looking around Ritničohkka's north-western slope and any patches of snow and ice. An unpleasant surprise awaited us upon our return to the cabins. We had left the door of the storage shed wedged open with a piece of wood as the lock was not fully functioning. Now, though, the strong gusts of wind had slammed it shut, breaking the lock for good and sealing the door tight. We had only little firewood left in the main cabin, and the storm did not show any signs of dying. The idea of being stuck in the cold, moist and mouldy cabin for three more nights grew increasingly disturbing. Even more disturbing was the forecast, accessible via the solitary mobile hotspot outside, which predicted a high probability of thunder for the next day. Even after the worst weather, there didn't seem to be any prospects for the helicopter to make it back to Ritničohkka for at least a week.

In light of the news and events, our cabin grew increasingly disturbing. The mould on the walls provided another tool for reflecting on our relations with Ritničohkka as a weird place. A cabin in the wilderness is a common motif in a subgenre of—especially American—horror films exemplified by the likes of *The Evil Dead* (1981) and *The Cabin in the Woods* (2011). A key theme is a cabin surrounded by an unknown, frightening, threatening and ultimately hostile environment represented by the forest, which either harbours a monster or is one itself (Whitaker 2020). While our sentiments about our cabin and its surroundings were not quite the same as in cabin horror fiction and the associated American "frontier" imagination (Whitaker 2020), we nonetheless felt very uneasy in and about this dank, cold and rundown cabin polluted with decades of mildew.

Our uneasiness was, in part, due to us being so prominently exposed to the difference and otherness of the relationally constituted "openness" of reality (which resonated with the openness of the physical landscape), which was permeated with invisible life and a sense of more-than-human sentience, awareness and agency. The cabin on Ritničohkka combined a sense of isolation with the feeling of "not being alone". But we were unable to pinpoint any tangible reason for this vague feeling of potential living presences of a more-than-human kind. The feeling resembled that of "evil stars" gazing on unaware characters in Lovecraft's stories. Whitaker points out that, "The unseen monster does not need to be literally invisible, but rather exists predominantly offscreen, or is never given a true form. Bizarrely enough, it is film where many of the most notorious unseen monsters can be found" (Whitaker 2020: 89). As for our experiences on Ritničohkka, we became more aware than ever before that knowing a relational world theoretically and intellectually is quite different from viscerally feeling and experiencing it.

The other source of weirdness as regards the cabin stemmed from a dissonance between what a cabin represents and how it felt. A cabin—like a house—is a distinctively human thing, meant to provide shelter and safety for fragile human bodies out in the wild. But the cabin on Ritničohkka, such as it was, contradicted these assumptions. Thus, rather than curtailing the strangeness of this land, the cabin amplified it. This is also why houses and architectural spaces play such an important role in Lovecraft's and other horror writers' fiction (Evans 2004, 2005). But again: experiencing this disturbing feeling oneself—and in a remote place—is quite different from the feelings produced by fiction. In a sense, our cabin worked against its intended and assumed function, as materially expressed by its constant decay and myriad micro-organisms and processes associated with ruination. Thus, the cabin itself became something like a wormhole or portal to non-human and otherworldly domains—or, at the very least, it tangibly manifested their existence and entanglement with the this-worldly. Our cabin was not a safe haven in the complex, unpredictable and weird relational reality of the High North but instead a monstrous being in itself. It was a place of latent existential horror in a strange and unsettling land that was somehow a sea of rocks and a mountain among mountain beings.

In essence, then, the cabin also manifested the central principle of uncertainty and open-endedness that characterizes the northern world as a relationally constituted reality. The cabin with its specific material properties contributed to taking us deeper into that reality instead of isolating us from it. When we first saw the cabin from the outside, it seemed sturdy and in a fairly good shape. However, entering it revealed a dark, disordered and unsettling microcosm, which underlined that in a relational world things are not always what they initially appear to be. As for the implications for archaeological fieldwork, this also posed an important question: if things are not necessarily what they seem, what should we document and pay attention to when fielding a weird land like this? What are meaningful features, aspects and dimensions of landscapes and things when they do not have fixed and stable identities? While there are no simple answers to these questions, the cabin, for instance, should be regarded both as a distinctively human thing and simultaneously a rabbit hole leading to otherworlds. Indeed, the latter aspect is perhaps not so different, in figurative terms anyway, from the revelation in *The Cabin in the Woods*, where—spoiler alert—the eponymous cabin turns out to be a gateway to a monstrous underground maze.

Reading up on the psychological dimensions of cabin horror, we discovered that "Cabin horror is a genre of repression and transgression; the isolated nature of the cabin in the woods renders it the perfect location to release one's repressed desires, in much the same way as it might be cathartic for a viewer to watch a horror movie and see said desires released on screen" (Lucas 2021: 51). As for the cabin on Ritničohkka, it certainly features as a place of transgression and perhaps, if somewhat less obviously, is also associated with repression. The latter is not so much about repressed individual desires, however, but about how the modern West has repressed chaotic, unpredictable and uncontrollable aspects of reality in the desire to subject it to human order and control. In contrast, relational thinking readily accepts those aspects as part and parcel of the world that people co-inhabit with multiple more-than-human beings and forces.

After the fieldwork, we wanted to explicitly communicate with a more-than-human-entity about the kinds of encounters that we had on Ritničohkka. The best we could come up with was to ask questions and converse with the artificial intelligence large language model ChatGPT (version GPT-3), which was trending immediately after being released for public use. Our hope was for a non-human conversant to illuminate new insights in ourselves. We took numerous different approaches to question it, prompting the AI to tell us, in various queries, both informatively and poetically, about the cosmological significance of cabins in different cultures and mythologies, as well as to freely "envision" cabins as portals to other dimensions of reality, for instance. While some parts of this communication were interesting, our non-human partner-in-dialogue, being the early language model that it was, could not really provide particularly useful or insightful perspectives on the specific matter of the cosmological significance of cabins. However, interestingly, ChatGPT happily "hallucinated" some replies to our queries, as it has become notorious for doing (cf. Seitsonen and ChatGPT 2023), telling us, for instance, that in Norse mythology the creation of the cosmos started with Odin building a cabin—a creation myth not told by any other source.

We will further consult ChatGPT briefly later in the book, but for now we can conclude that dwellings have indeed been cosmologically important in northern cultures and conceived as microcosms of reality (Seitsonen and Fjellström 2022; Herva and Lahelma 2020; Westerdahl 2002; Anderson 2013). Dwellings can be seen as miniature representations of the larger world and its varying structures, values and meanings. A dwelling always reflects in myriad ways the culture, history, values and identities of its inhabitants and users, as well as their relationships with each other and with the environment. Dwelling is also the human condition of being in, and an ongoing process of engagement with and experiencing, our environment, in a constant process of becoming as we encounter the world (Ingold 2000; Seamon 1993).

References

Äikäs, T. 2015. *From Boulders to Fells: Sacred Places in the Sámi Ritual Landscape*. Helsinki: The Archaeological Society of Finland.
Anderson, D.G. 2013. Home, Hearth and Household in the Circumpolar North. In *About the Hearth: Perspectives on the Home, Hearth and Household in the Circumpolar North*, ed. D.G. Anderson, R.P. Wishart, and V. Vaté, 262–282. New York: Berghahn.
Aupers, S. 2009. "The Force Is Great": Enchantment and Magic in Silicon Valley. *Masaryk University Journal of Law and Technology* 3 (1): 153–173.

Bradić, M. 2019. Towards a Poetics of Weird Biology: Strange Lives of Nonhuman Organisms in Literature. *Pulse: The Journal of Science and Culture* 6 [online]: 1–22.
Brown, B., and M. Chalmers. 2003. Tourism and Mobile Technology. In K. Kuutti, E.H. Karsten, G. Fitzpatrick, P. Dourish, and K. Schmidt (Eds.), *Proceedings of the Eighth Conference on European Conference on Computer-Supported Cooperative Work*, 335–354. Helsinki: Kluwer Academic.
Dickinson, J.E., J.F. Hibbert, and V. Filimonau. 2016. Mobile Technology and the Tourist Experience: (Dis)Connection at the Campsite. *Tourism Management* 57: 193–201. https://doi.org/10.1016/j.tourman.2016.06.005.
Douglas, M. 1966. *Purity and Danger: An Analysis of the Concepts of Pollution and Taboo.* New York: Routledge.
Evans, T. 2004. Tradition and Illusion: Antiquarianism, Tourism and Horror in H.P. Lovecraft. *Extrapolation* 45 (2): 176–195.
Evans, T. 2005. A Last Defense Against the Dark: Folklore, Horror, and the Uses of Tradition in the Works of HP Lovecraft. *Journal of Folklore Research* 42 (1): 99–135.
Fisher, M. 2016. *The Weird and the Eerie.* London: Repeater.
Fonneland, T., and T. Äikäs. 2023. *Shamanic Materialities in Nordic Climates. Cambridge Elements.* Cambridge: Cambridge University Press. https://doi.org/10.1017/9781009376396.
Herva, V.-P. 2010. Buildings as Persons: Relationality and the Life of Buildings in a Northern Periphery of Early Modern Sweden. *Antiquity* 84 (324): 440–452.
Herva, V.-P., and A. Lahelma. 2020. *Northern Archaeology and Cosmology: A Relational View.* Abingdon: Routledge.
Herva, V.-P., and O. Seitsonen. 2020. The Haunting and Blessing of Kankiniemi: Coping with the Ghosts of the Second World War in Northernmost Finland. In *Entangled Beliefs and Rituals*, ed. T. Äikäs and S. Lipkin, 225–235. Helsinki: The Archaeological Society of Finland.
Herva, V.-P., and T. Ylimaunu. 2009. Folk Beliefs, Special Deposits, and Engagement with the Environment in Early Modern Northern Finland. *Journal of Anthropological Archaeology* 28 (2): 234–243.
Ingold, T. 2000. *The Perception of the Environment: Essays on Livelihood, Dwelling and Skill.* London: Routledge.
Korpijaakko, K. 1989. *Saamelaisten oikeusasemasta Ruotsi-Suomessa: Oikeushistoriallinen tutkimus Länsi-Pohjan Lapin maankäyttöoloista ja -oikeuksista ennen 1700-luvun puoliväliä.* Helsinki: Lakimiesliiton kustannus.
Lahelma, A. 2007. "On the Back of a Blue Elk": Recent Ethnohistorical Sources and "Ambiguous" Stone Age Rock Art at Pyhänpää, Central Finland. *Norwegian Archaeological Review* 40 (2): 113–137.

Lang, T., and W.T. Borrle. 2021. Wilderness Solitude in the 21st Century: A Release from Digital Connectivity. *Science & Research* 27 (3). https://ijw.org/wilderness-solitude-arhertog-release-from-digital-connectivity/.

Lea, R. V. 2001. The Performance of Control and the Control of Performance: Towards the Social Anthropology of Defecation. PhD thesis, Brunel University, London.

Lucas, K. 2021. Flipping the Castle: Evolution of Gothic Spaces in the Domestic Sphere. BA thesis, The College of William and Mary.

Muhonen, T. 2008. Something Old, Something New: Excursions into Finnish Sacrificial Cairns. *Temenos* 44 (2): 293–346.

Okkonen, J. 2003. Jättiläisen hautoja ja hirveitä kiviröykkiöitä: Pohjanmaan muinaisten kivirakennelmien arkeologiaa. PhD thesis, University of Oulu, Oulu.

Pearce, P., and U. Gretzel. 2012. Tourism in Technology Dead Zones: Documenting Experiential Dimensions. *International Journal of Tourism Sciences* 12 (2): 1–20.

Rainio, R., and E. Hytönen-Ng. 2023. Ringing Tone and Drumming Sages in the Crevice Cave of Pirunkirkko, Koli, Finland. *Open Archaeology* 9 (1). https://doi.org/10.1515/opar-2022

Salmi, A.-K., and O. Seitsonen. 2022. Effects of Reindeer Domestication on Society and Religion. In *Domestication in Action: The Anthropology and Archaeology of Reindeer Domestication in Fennoscandia*, ed. A.-K. Salmi, 215–248. Cham: Springer.

Salmi, A.-K., T. Äikäs, M. Fjellström, and M. Spangen. 2015. Animal Offerings at Sámi Offering Site Unna Saiva: Changing Religious Practices and Human-Animal Relationships. *Journal of Anthropological Archaeology* 40: 10–22.

Schultze, U., M. Aanestad, M. Mähring, C. Østerlund, and K. Riemer, eds. 2018. *Living with Monsters? Social Implications of Algorithmic Phenomena, Hybrid Agency, and the Performativity of Technology*. Cham: Springer.

Seamon, D. 1993. *Dwelling, Seeing, and Designing: Toward a Phenomenological Ecology*. Albany: State University of New York Press.

Seitsonen, O. 2021. *Archaeologies of Hitler's Arctic War. Heritage of the Second World War German Military Presence in Finnish Lapland*. Abingdon: Routledge.

Seitsonen, O., and ChatGPT. 2023. Tekoälyn kanssa iltapäiväkahvilla. *Muinaistutkija* 1 (2023): 69–79.

Seitsonen, O., and M. Fjellström. 2022. Habitation Sites and Herding Landscapes. In *Domestication in Action: The Anthropology and Archaeology of Reindeer Domestication in Fennoscandia*, ed. A.-K. Salmi, 153–186. Cham: Springer.

Seitsonen, O., and V.-P. Herva. 2011. Forgotten in the Wilderness: WWII PoW Camps in Finnish Lapland. In *Archaeologies of Internment*, ed. A. Myers and G. Moshenska, 171–190. New York: Springer.

Seitsonen, O., V.-P. Herva, K. Nordqvist, A. Herva, and S. Seitsonen. 2017. A Military Camp in the Middle of Nowhere: Mobilities, Dislocation and the Archaeology of a Second World War German Military Base in Finnish Lapland. *Journal of Conflict Archaeology* 12 (1): 3–28.

Tanti, A., and D. Buhalis. 2017. The Influences and Consequences of Being Digitally Connected and/or Disconnected to Travellers. *Information Technology and Tourism* 17: 121–141. https://doi.org/10.1007/s40558-017-0081-8.

Tsing, A.L. 2015. *The Mushroom at the End of the World: On the Possibility of Life in Capitalist Ruins*. Princeton: Princeton University Press.

Tuovinen, T. 2002. *The Burial Cairns and the Landscape in the Archipelago of Åboland, SW Finland, in the Bronze Age and the Iron Age*. Oulu: University of Oulu.

Vesajoki, H. 2021. *Pohjois-Karjalan kivet ja kallio kertovat*. Keuruu: Self-Published.

Westerdahl, C. 2002. The Heart of Hearths: Some Reflections on the Significance of Hearths in Nature, Culture and in Human Memory. *Current Swedish Archaeology* 10: 179–198.

Westerdahl, C. 2005. Seal on Land, Elk at Sea: Notes on and Applications of the Ritual Landscape at the Seaboard. *International Journal of Nautical Archaeology* 34 (1): 2–23.

Whitaker, B. 2020. The Forest Is Not What It Seems: An Ecocritical Study of American Horror Films. MA thesis, Middle Tennessee State University.

Zijlstra, J., and I. van Liempt. 2017. Smart(Phone) Travelling: Understanding the Use and Impact of Mobile Technology on Irregular Migration Journeys. *International Journal of Migration and Border Studies* 3 (2/3): 174. https://doi.org/10.1504/IJMBS.2017.083245.

Open Access This chapter is licensed under the terms of the Creative Commons Attribution 4.0 International License (http://creativecommons.org/licenses/by/4.0/), which permits use, sharing, adaptation, distribution and reproduction in any medium or format, as long as you give appropriate credit to the original author(s) and the source, provide a link to the Creative Commons license and indicate if changes were made.

The images or other third party material in this chapter are included in the chapter's Creative Commons license, unless indicated otherwise in a credit line to the material. If material is not included in the chapter's Creative Commons license and your intended use is not permitted by statutory regulation or exceeds the permitted use, you will need to obtain permission directly from the copyright holder.

CHAPTER 6

The Fjell in the Cloud

> This is not a place for the living
> the desolation of dying ice
> The drip, a clock
> a future and past
> From a mighty proud glacier
> to a sad pathetic pond
> And the lifeless grey stones
> in alien angles protrude
> An unsettled landscape
> which impenetrable mists entomb
> —"*Poe*tic Pastiche" #3, Aki Hakonen

6.1 "This Is Where You Find Cthulhu"

On the second day around noon, as the clouds had shifted just enough to provide fleeting glimpses of the world beyond them—enough to show that only the top of Ritničohkka was submerged—we ventured out to continue the survey. We first navigated in the cloud for some time, which made the fjell look and feel very peculiar. Unable to see the horizon in any direction, only the grey rocks and ground in our immediate vicinity offered points of reference, which disappeared as we walked further in the grey cloud. Occasional rock formations loomed in the mist. The scene was reminiscent, yet again, of a horror or sci-fi film, as we were experientially definitely "someplace else" outside the familiar everyday environment.

Intriguingly, when we posted pictures on social media later on, somebody commented on a picture showing us walking in the cloud among the rocks—unaware of our own reflections—that "this is where you find Cthulhu" (Fig. 6.1). Clearly, we were onto something in our designation of the place as Lovecraftian.

Indeed, one can readily imagine a resonance between the photograph and Lovecraft's description, in the short story *The Call of Cthulhu* (1928), of stranded sailors climbing the momentarily surfaced sunken city of R'lyeh located somewhere in the Pacific Ocean, composed of weird rocks and featuring an incomprehensible non-Euclidean geometry. This reference also resonates with the reflections of the vast boulder field as a weird sea.

Ritničohkka-in-a-cloud was very disorienting, as everyone who has tried orientating themselves in similar conditions knows. Our general directions were guided by compass and small heaps of rocks that we set up along our way, although due to the near lack of visibility we had to rely on a GPS navigator. Walking in the cloud afforded a curious combination of liberation, confinement, excitement, restlessness and calm. There was even a spiritual dimension to it, something similar to altered states of consciousness, walking inside a grey bubble with nearby yet indiscernible

Fig. 6.1 A monstrous landscape and an imprisonment of perception in the mist, at once awe-inspiring and horrific

boundaries and incomprehensible features (comparable to the feeling of skiing in a whiteout). This forms a different, dream- and nightmare-like vision of the surrounding reality, prompted by the qualities of the landscape itself—its transcendental beauty and its penetrating but hard-to-describe weirdness. Such modes of perception and awareness of oneself in relation to the surrounding world are, in fact, integral to the traditional northern, relational ways of knowing the world (Herva and Lahelma 2020).

There was an amplified uncertainty about the form and characteristics of the things we saw looming in the mist. Shadowy forms coming and going, materializing into view and disappearing, shifting, vibrated with a sense of mystery. Things in the world presented themselves differently from a clear-eyed view, hence also underlining the changing, unstable and mutable nature of this landscape under—or entangled with—the more readily accessible yet potentially deceptive surface-world of seemingly lifeless rocks (see also Ahola and Lassila 2022).

The mist effectively hid the landscape but simultaneously opened up new perspectives and aspects of the same landscape. It hinted at the multitude and multidimensionality of the land; so, paradoxically, the cloud not only hid but also revealed the different sides and layers of this world. It also suggested something of our place within that closed-up land.

6.2 All the Shades of Grey

Glimpses of a richer and deeper reality beneath the surface of things notwithstanding, Ritničohkka was distinctly grey on our second day—so much so that it occasionally felt as if colours had been sucked from this environment, and even from our own rain clothes, replaced by a monochrome reality. This provoked a mirroring of states of mind, from the mysterious to gloominess and calm which in turn mediated our view of and relationships with the landscape—especially powerfully so when we were exploring the lingering remains of melting ice and what it represents. Although we were not consciously analysing our thoughts or emotions as we went along through the cloud, concentrating on not slipping or tripping on the wet and slick rocks, we later came across a reflection of grey that "clicked" with our general mindscape of that day. As Pétursdóttir (2018: 4) writes:

"Grey is the fate of color at twilight" (Cohen 2013, 270). Grey's immediate association isn't hopeful and optimistic, but dull, lifeless, deprived. "A grey ecology", as Cohen explains, "might therefore seem a moribund realm, an expanse of slow loss, wanness, and withdrawal, a graveyard space of mourning" (ibid.). Grey is the shade of sickness. It may be the hue of Anthropocene, of anxiety, regret and bleakness. The lowering light preceding the apocalypse.

But grey is also a liminal shade, withdrawn in its prosaic patina it allows the elements to rest. It flattens the plane, soothes contrasts and brings differences closer. Grey brings us another landscape—not one that is colourless, dull and lifeless, but a landscape tinted by tranquillity, enabling the eye to glide unhindered between elements of different kind. Grey pacifies our preconceptions and tones down our definitions.

Grey, then, has more to it than might initially seem on the surface. Like a boulder field, it first looks monotonous and lifeless, but it both hides and reveals a richness of underlying things and dynamics, which is something that we experienced on Ritničohkka (Fig. 6.2). More generally, walking through the wet, dark and impenetrable cloud was a reminder of how fieldwork in certain types of environments involves much more than data collecting. It can produce broadly spiritual experiences and insights towards ourselves and how we relate to the surrounding world, including a new awareness of its entangled weirdness. In other words, doing fieldwork in a strange, alien and remote land can yield profound and transformative experiences, including encounters with more-than-human entities that afford a restructuring of our connections with the world and the discovery of novel connections, which is what a "magical consciousness" is ultimately about (Greenwood 2009).

Our fielding on Ritničohkka—with our enhanced bodily-cognitive awareness of and openness to the land—underlined the significance of situationality, which is central to relational knowing and being in the world. We obviously cannot perceive, experience and relate to the land similarly as people in the past have done. However, our heightened awareness of myriad tangible and intangible features and dimensions of the landscape "calibrated" us to detect and appreciate the multiplicity, richness and relevance of those dimensions. While we documented our finds, mainly reindeer bones, in a conventional manner, it is important to recognize that walking itself is also a modality of knowing the world and our relationships with it (see de Certeau 1984: 91–110; Gros 2023), even if

Fig. 6.2 Grey hides and reveals the last of the Ritničohkka snow and ice patches

it is not readily obvious how the knowledge generated this way can or should be documented in the context of archaeological fieldwork.

Our fieldwork was prospective in nature, with the general objective of surveying the area of the snowfield to detect any finds and features of archaeological interest. But our movement in the landscape was relatively unplanned and largely guided by the land itself. We moved with the landscape in ways that it enabled us to, rather than going against it, for instance, by walking in direct parallel transects across the land—as is a typical way of surveying in the archaeological fieldwork tradition that strives for statistical coverage to "conquer" an untamed and chaotic landscape. There was an element of "drifting" to our fielding, which has similarities to psycho-geographical walking, or the aforementioned *dérive* (Debord 1956). Significantly, such "drifting" involves shifting states of consciousness, which indeed was an important element in our "tuning into" Ritničohkka as a relationally constituted and known entity. "Walking", as Graves-Brown and Schofield point out, "highlights terrain, geology, the impact of changing weather, the sounds and smells of landscape, and its history. [...] [Walkers] feel the landscape and come close to becoming a part of it—finding place and often also finding themselves through this slow and intimate form of encounter" (Graves-Brown and Schofield 2020: 64–67).

6.3 Disorientation and Labyrinths of Stone

Ritničohkka is a rather poorly known fjell in Finland, unlike Háldi next to it, which has gained a mythical standing in Finnish mindscapes. As mentioned previously, Háldi features in the local Sámi mindscape as the mythified mother of reindeer (Seitsonen and Viljanmaa 2021). Another equally mythical fjell, Sána, is located next to the village of Kilpisjärvi, around which we had done several years of exploratory work before the Ritničohkka expedition. We were passively aware of two large stone labyrinths on top of Sána, right below the peak of the fjell, but we only came to think about them in 2020, when we took a group of students in a fieldwork course there (see Paphitis et al. 2021). We knew that the labyrinths are recent, evidently built sometime in the early 2000s, although nobody seems to know who constructed them. Locals suspect it was bioartists working in Kilpisjärvi, but they deny having made them. Curiously, on the same 2020 trip, a student found another analogous design, a spiral-dominated piece of rock art, which under closer investigation turned out to be a work from the 1980s by Japanese-American artist Ken Hiratsuka. A little later, we discovered that there is another related piece by the same artist at the famous-in-some-circles, pseudo-archaeological site known as the "Temple of Lemminkäinen" in the very south of Finland (see further Herva et al. 2024).

All these connections spurred us to explore questions of stone labyrinths, or "Troy Towns". These number several hundred in Finland and Sweden, mainly constructed in maritime landscapes of the Baltic Sea, but also found on the Arctic coast of Norway and the White Sea in Russia (see Aspelin 1877; Uino 2003; Westerdahl 1991). While there are no labyrinths on Ritničohkka (as far as we know), labyrinths provided one conceptual tool—or an echo chamber—for our attempts to make sense of the Ritničohkka landscape and how it affected us. The labyrinth of classical Greek myth is famously the prison of a monster, the Minotaur, but it has a host of other characteristics that afford a tool for thinking about, or a point of comparison to, our experiences on Ritničohkka. The labyrinth as a visual and spatial design has fascinated people for thousands of years, and much has been written about the idea, design, metaphor and material manifestations of the labyrinth from different perspectives and in various contexts (e.g. Doob 1990; Artress 1995; Kern 2000; McCullough 2005; Conway 2011; Harris 2014; Schmidt di Friedberg 2017).

Ritničohkka is characteristically a "natural" place, but our experience of and engagement with the fjell has some similarities to what labyrinths are and do. The labyrinth is confusing by design and is "activated" by walking it. The activity of walking the labyrinth can have spiritual and transformative effects (Artress 1995). The labyrinth, with its ambiguous and ambivalent character, is a fitting place for monsters and the monstrousness. This monstrousness is reflected in contemporary fantasy and horror, such as the films *Pan's Labyrinth* (2006), the aforementioned *Cabin in the Woods* (2011) and Denis Villeneuve's thriller *Prisoners* (2013). The labyrinth design dates back several millennia and circulated far and wide. It implies movement. And not just any movement, but the kind of movement that manipulates the sense of time, place and self. Our wandering and drifting on Ritničohkka had similar effects. It felt like the landscape guided us deeper into it, gradually revealing different layers of the fjell-world in a similar manner as a labyrinth takes its walker—in both the classical myth and modern experiments—into a liminal world outside ordinary reality. The liminal world is most often associated with the subterranean and involves dissolving boundaries between "inner" and "outer" worlds (e.g. Artress 1995; Harris 2014; Doner 2022). This dissolvement allows for connecting and restructuring perceptions and relationships between self and the world, which in turn is what magical consciousness is ultimately all about: an awareness of how things in the world are interconnected (see Greenwood 2009).

Walking on Ritničohkka was characterized by slow, careful and concentrated steps, particularly in the area of the former ice field, with attention alternating between steps, scanning the surroundings for finds and momentarily pausing to stare at the landscape. This rhythmic alternation of attention can be compared to the rhythmic turns and twists in walking a classical unicursal labyrinth, which in turn is one factor that induces the particular mental or spiritual effects of experiencing the labyrinth (Artress 1995; Doner 2022). The quintessential effect of the labyrinth is attained by moving literally and figuratively deeper into the maze, accessing an otherworldly reality. Disorientation can be considered an essential aspect of the weird (Lockhurst 2017). The disorientation that the labyrinth provokes can be related to the disorientation we experienced on Ritničohkka, further promoted by various environmental and material factors described throughout this book. Disorientation is a form of altered state of consciousness and relates to the otherworldly dimensions of lived reality. This notion that labyrinths are gateways is perhaps clearly illustrated, in the context of the Arctic, by the stone labyrinths constructed at Sámi burial sites on the Arctic Ocean coast in Norway (see Olsen 1991). These archaeological examples were the likely blueprints for the modern labyrinths built on top of Sána.

A labyrinth implies movement—and even a particular way of moving in and through a particular space—closely related to dancing. Movement or dance is reflected in the classical myths and Finnish/Swedish folklore associated with stone labyrinths, which apparently date primarily from the mediaeval and early modern period (Ikäheimo 2011; Niukkanen 2009). Among their common vernacular names is the Finnish-Swedish *jungfrudans* ("maiden's dance"). It has been speculated that the labyrinth design perhaps originates from a dance pattern or choreography, which readily takes us to the meaningfulness of bodily movement, and how the differing ways of moving in and engaging with a place or a landscape affect the ways we perceive ourselves and the surrounding world. Another suggestion is that they have been used in maritime weather magic, trapping the incorporeal ghosts of drowned sailors who were said to induce storms (Westerdahl 2014).

As for the cultural associations of the labyrinth in Minoan Crete, dancing appears to have been ceremonially important in Minoan Crete and associated with altered states of consciousness (e.g. Morris and Peatfield 2006, 2012). While our moving through the Ritničohkka landscapes was not formally choreographed, it nonetheless comprised a specific mode of moving and perceiving in relation to the landscape, particularly around the glacier with its challenging terrain, which called for and induced a mode of movement that made us particularly acutely aware of ourselves and our surroundings: enminded bodies slowly "dancing" with rocks and the landscape in circles reaching out from our basecamp. Although not specific to Ritničohkka, this was a very different way of moving from ordinary, everyday human-designed landscapes, where walking and moving tends to be more subconscious, linear and automatic or mechanical. It differed also from the way traditional archaeological surveys in forests or on the tundra unfold, where, even when you are more aware and conscious of your motion, you do not need to be so acutely aware and careful as when traversing a "post-glacial" landscape of loose, slippery, vegetationless rocks. The disorientation and acute awareness of physical motion even further underlined the "otherness" and otherworldliness of the place, especially since it looked so easily traversable from the deceitful panopticon of a hovering helicopter.

6.4 Manifest and Hidden Dimensions

In a fascinating study, Ahola and Lassila (2022) explored storytelling and performance in relation to a Mesolithic bone artefact from Finland. They considered, for instance, the possible use of this artefact as a puppet for shadow play. They documented the diversity of shapes that this three-dimensional artefact can cast, both intentionally and controllably and, with light from a living fire, unintentionally and uncontrollably. They speculate that

> the unpredictable interplay of light and shadow might, in fact, have been the key to the use of the object. In other words, the vague qualities of the artefact were not only experienced, but also orchestrated on purpose. From a relational perspective, the artefact would have had an agency of its own. [...] Indeed, it is possible that the Mesolithic people understood the shadowy forms of the item as the souls of the animals these forms represented. (Ahola and Lassila 2022: 180)

While it obviously cannot be "proven" that the meanings of the studied artefact involved this kind of dimension, Ahola and Lassila cite ethnographic information about northern hunter-gatherers and their concepts of "hidden sides" and "shadow souls". The basic idea applies also more generally in regard to both northern cultures and relational cosmologies: things and beings have open and multiple identities which manifest themselves differently and in different situations and conditions. These conditions encompass both external and internal factors, from, for example, lighting and weather conditions to particular frames of mind and modes of awareness. Consequently, too, in order to know the multiplicity of what things really "are" and how they are related to other things, it is necessary to "circulate" them (cf. Tilley 1999). This can be done both literally, with different ways of engaging them, and figuratively, by "seeing" and thinking about them from different perspectives and based on different worldviews. The same applies to places and landscapes—and, most pertinently here, as they presented themselves to us on Ritničohkka.

The nights spent in the Ritničohkka cabin were phantasmic even in comparison with nights in a regular wilderness cabin. Even though August nights in the North are luminous, with the sun dipping ever so slightly below the horizon, the dampness of the practically abandoned lodge seemed to repel light from the outside and suck it into its dark corners. It was as if its mould-ridden walls absorbed all excess light seeping through the dirty panes of the cabin's single small window. Meanwhile, the wood-burning stove would not warm the cabin, even though we maintained a fire in it for hours and hours, exhausting our fast depleting firewood. The cabin itself made constant noises—creaking, even sounds that reminded us of breathing—as the wind gusted around us, with the sounds amplified during the night.

Light and shadow are integral to how we perceive forms. As such, different light conditions have myriad impacts on how we sense place and space (e.g. Dowd and Hensey 2016; Papadopoulos and Moyes 2022). Following the cue from Ahola's and Lassila's (2022) research, and considering places in terms of relational ontologies and epistemologies, things do not simply "look" or "appear" different depending on a wide range of internal and external factors, but rather show or reveal different—albeit interlocked and intertwined—"layers" of reality, its different sides or faces, which unfold under different conditions and from different perspectives.

This is not, however, limited only to visual engagement with the world but applies also to other ways of sensing, knowing and connecting with its tangible and intangible aspects. For an interesting illustration, consider Marila's (2020) experimentation with materiality, sound and human-environment relations. The starting point for this experimental work in art/archaeology was the phenomenon known as "The Hum", a somewhat elusive and weird phenomenon which a few per cent of the global population experience and suffer from. It manifests itself as a low frequency droning sound, often associated with specific areas, and its source or cause is unknown; it might be due to external causes (such as electronic transformers, mobile phone networks or military radio activity) or inner causes, in which case the hum would presumably be a special type of tinnitus (Marila 2020).

Marila (2020: 107–109) briefly surveys some intriguing hypotheses or theories that researchers have proposed on the hum, including the idea that "a small percentage of the population may have a bodily ability to turn radio waves into auditory experience" and that "nerve impulses are not electrical but sonic in nature." More outlandishly—literally—it is arguably possible that the hum originates in planetary oscillations or some other cosmic phenomena, and would therefore literally be out of this world. After all, some people have long claimed to be able to hear the northern lights, and recently it was discovered that they do indeed emit sound (Marila 2020).

As Marila (2020: 109) points out, "There is much about the human body and the senses and their overlap, therefore, that remains poorly understood." Moreover, "Just as our bodies may be able to pick up radio waves and turn them into sound perception, so also may our bodies be able to 'hear' the gravitational waves emanating from, for example, black holes" (Marila 2020: 109, citing Abbott 2016). Synaesthesia is a medically known example of this phenomenon, presenting a "union of senses", where perceptual stimulation of one sensory or cognitive pathway causes involuntary experiences in another, unrelated sensory or cognitive field, such as seeing words as different shapes in your mind, hearing colours or tasting shapes (e.g. Cytowic 2018). Based on clinical studies, it has been estimated that about four per cent of the population experience this phenomenon which appears to be considerably more common among people on the autism spectrum (Baron-Cohen et al. 2013).

Inspired by the mysterious hum phenomenon, Marila observed that it "highlights the human body as an archaeological artefact in which soundscapes become felt, accumulated, and ossified". Marila consequently embarked on an art/archaeology experiment, The Hum, using redundant bone materials from archaeological contexts, armed with "the realization that acoustic experiences become stratified in the human body over long periods of time. Sounds stick to the body in ways that are fundamental to the body's presence" (Marila 2020: 109). In collaboration with the sound artist Tony Sikström, Marila had the animal bones pulverized, creating acoustic installations from the powder. In effect, this artistic/ aesthetic experiment revealed properties of archaeological bone finds that are "there" in the bones and resonate with external acoustic stimuli.

Whether or not this has any direct bearing on understanding more about the specific bone finds and how past people engaged with them (and the animals that they represent) is irrelevant here. What matters, instead, is that there are layers to things—or aspects to their "identities" and being in the world—that can be revealed in many different ways, both scientific and creative. Moreover, the process of engaging creatively with archaeological finds brought together and opened up unobvious and unexpected dimensions of things and how they are entangled with other things.

The Hum sets out to demonstrate that, whereas weird things happen in the human body when information (like radio waves) is transduced into another form of information (such as the auditory experience of those radio waves), the same holds for mechanical apparatuses when they interpret their surroundings. Instead of following a clear logic of mechanistic causation, technological interpretation is faced with the same ambiguities and uncertainties as are faced by human perception. (Marila 2020: 111)

6.5 Enminded Bodies and Other Instruments for Accessing Invisible Worlds

Several philosophers have argued that cognition is not something that happens in the brain, but emerges from interaction with our surroundings, often mediated by objects. For example, Merleau-Ponty (1962) describes our lived and experienced body as a purposeful body-subject that makes meaning in the world and enables our pre-reflexive, subconscious being in the world as well as engagement with the surrounding world. The things that we use and that surround us extend and delimit our perceptual-cognitive-bodily capabilities, essentially making humans into extended "human-object hybrids" (Edensor 2005). This idea of us and our possessions forming extended hybrid beings is profound. A walking stick helps to balance hikers in difficult terrain, and enables blind persons to "see" the environment differently from not having a stick in their hand. Eyeglasses can enhance vision, transforming the world for the beholder. Similarly, a traditional old-school phone enables talking with people far away, while a smartphone enables interaction with a cyberworld, expanding as well as contracting the senses and the experienced world.

But modifications of perception, cognition and communication with the world are not limited to these kinds of concrete examples and cases. There are multiple ways for people to tune into different frequencies of reality, or realign themselves towards it, so as to perceive and connect with reality differently. That has been the thinking in various traditions and cultures, including in Western esotericism, with its myriad entanglements with scientific thinking and technology. During the nineteenth century, occult revival, mediums and séances became popular and were represented as a way of communicating with the spirits of the dead (e.g. Dixon 2007). Likewise, spirit photography emerged basically as soon as photography was invented (Harvey 2007).

Séances, spirit photography and a host of other occult, esoteric or related philosophies and practices are often readily dismissed as nonsense and hoaxes. But this would be too hasty in that, considered in cultural terms, they reveal much about the long-standing intuitive sense that there is more to the world and its workings than is readily obvious on the surface of things. And this, of course, is what science and technology repeatedly demonstrate in revealing previously unknown and not directly perceivable things, structures, dimensions and connections between various things in the world.

In closer contextual terms, mediumistic practices and spirit photography, for instance, make cultural sense in that they resonate with the new technologies of their time and their broader implications. Thus, the invention of the telegram and telephone made it possible to communicate in real time over a distance, so the idea of communicating with the dead—by using the body and mind as a connective instrument—was just an extension of this idea and principle (Dixon 2007). Likewise, the camera "sees" the world differently from humans and enables, for instance, picking up and documenting things and details that might otherwise go unnoticed, so the idea that a camera could also see other dimensions of existence—spirits, in this case—again makes cultural and contextual sense (Harvey 2007). When the atomic theory was emerging, scientific and occult thinking became intertwined for a while, and the human body and mind featured as a potential way of connecting with invisible worlds in such forms as "clairvoyant chemistry" (Morrison 2007).

In the context of the circumpolar North, shamanism comprises an example of similar ideas in which certain bodily and mind-altering techniques enable accessing and interacting with planes or dimensions of existence that are ordinarily hidden or invisible. This is illustrated by the tiered shamanistic cosmology, which could be described as a layered reality. These different "layers" are not necessarily "geometrically" superimposed but comprise, rather, an entangled non-linear reality (e.g. Herva and Lahelma 2020). In this reality, a properly knowledgeable person (a shaman, or *noaidi* in the Sámi language, or *tietäjä* in Finnish, literally a "knower", that is, a seer; a witch, from Old English m. *wicc/a*, f. -*e*) can travel, in altered states of consciousness, into other dimensions of existence in search of special knowledge towards working out some problem manifested in the "this-worldly" plane of existence. In principle, and for heuristic purposes, this can be compared to what instruments such as the telescope and microscope do: they afford seeing things in this reality that cannot be perceived with the bare eye. It is similar to what magic is all about: employing various techniques to manipulate perception and consciousness towards "seeing" and becoming conscious of the surrounding reality and one's place in and entanglements with it from a differing perspective (e.g. Baird 2004; Greenwood 2009).

Traditional rationalist thinking conceives altered states of perception and consciousness as purely internal mental states that do not reveal anything about the world at large. Within relational thinking—and the idea of "magical consciousness" (e.g. Greenwood 2009)—such altered states can be understood rather differently and seen as producing insights into our being in relation to the surrounding world, thereby producing knowledge about some aspects of our lived reality. In effect, and in more concrete terms, this is similar to how incorrectly exposed photography can reveal aspects of the environment and its dynamics that are invisible in correctly exposed photographs or to the bare eye—again, a way of seeing the world differently.

As for altered states of consciousness, it can be argued that they do not, in fact, distort the otherwise correct perception of the world, but rather reveal dimensions of reality beneath the surface of things as we ordinarily perceive them (e.g. Luke 2010). Today, we are used to thinking of the brain as a computer, but this metaphor is merely a product of our time rather than an actual fact. Indeed, some argue that the "brain is not a computer [but] a transducer", an organ that assembles and converts signals (Epstein 2021).

These ideas bring us back to weird fiction again. Lovecraft, for instance, builds on the notion that the world as we ordinarily perceive it is but an illusion or thin veil beneath which lies the reality of a roaring chaos of unimaginable cosmic horrors. A non-fiction variant of this same notion came to haunt the renowned and troubled physicist Wolfgang Pauli (1900–1958). Immersing himself deep in the new reality of quantum physics, he, as told by author Tobias Hürter (2022), began seeing disturbing nightmares of inexplicable geometries and symbols. In his work, Pauli became especially transfixed on explaining why the divider for the fine-structure constant, which prevents atomic structures from collapsing, is 137. In 1931–1934, he was treated by psychoanalyst Carl Jung, who recognized Pauli's dream symbols as alchemical signs, and came to note that in Hebrew numerology 137 is the combined value of the letters in the esoteric word Kabbalah. (During Pauli's psychoanalysis by Jung, half-way around the world, Lovecraft, an obscure fringe writer, wrote the 1933 short story *The Dreams in the Witch House*, in which Walter Gilman, a student of physics, is haunted by inexplicable geometric nightmares. Jung began publishing Pauli's dreams beginning in 1935. Lovecraft died only two years later, in 1937.) Jung and Pauli later collaborated to develop the concept of synchronicity, concerning meaningful coincidences that do not seem to share a cause and effect (Halpern 2020). Decades later: "[…] Pauli is hospitalised with severe stomach pains. When he sees his room number at the Red Cross hospital, he exclaims 'It's 137! I won't be leaving this room alive.' Ten days later, Wolfgang Pauli is dead" (Hürter 2022).

Searching for deeper meaning in these seemingly disconnected associations is far-fetched, but it serves to illustrate the broader cultural context of relativism, synchronicity and weird fiction that affected our experiences on Ritničohkka—our embodied perceptions and sense of the place. This does not, of course, suggest that there is *literally* a reality in a cosmic scheme of Lovecraftian monsters, or of visions as experienced in altered states of consciousness, beneath the mundane reality that we ordinarily perceive. Rather, this discussion simply highlights that there is more to lived reality than objectivist approaches to landscapes or enminded-bodies-in-the-landscape can readily grasp and describe—the lived and experienced world is layered in complex ways, in both a literal and figurative sense. Reality itself is largely relative to the one who experiences it.

The world as we humans experience it is a weird place. And, in order to make sense of it as best we can, we need to account for that tangible and intangible weirdness in diverse forms. This involves accepting all the meaningful, real or imagined aspects and dimensions of reality that there are in the experienced world, and mediating our relationships with them. This is what thinking through the Anthropocene and its ontologically varying monsters ultimately seeks to do. Also, much of our lives unfolds in relation to the unknown in diverse forms and to things we are not actively aware of. Consequently, it requires a conscious effort to become more aware—with various conceptual and material instruments, our own body-subjects included—of the various things that mediate our relationships with places and landscapes. Awareness leads to a better appreciation of *how* we are in the world and relate to it. This *how* takes the form of emotions, feelings and affects, shedding light, in turn, on how different visible and invisible—tangible and intangible—constituents of the surrounding world affect our being and thinking in the world.

6.6 Air and Smells as Intangible Artefacts

Air tends to be perceived as emptiness in everyday vernacular thinking, but in reality, of course, this is not what air is. Besides its physical molecular structure, air is perceivable in many ways, especially as a carrier of scents. In cities and towns, as well as in the countryside, different scents are all around, indicating restaurants, garbage bins, bakeries and sewers. Car emissions and freshly cut grass, rotting autumn leaves mixed with dog faeces and pig manure produce an abundant multisensory scenery of smells. Industrial towns, especially places with paper mills, are known for their strong and recognizable odours.

It is commonly thought that for humans, compared to many other animals, scents represent a largely hidden world. The human sense of smell may have been slowly diminished, possibly even by our "domestication" to an agricultural and urban world, and, more recently, due to pollution (Hoover 2018; Li et al. 2022; cf. McGann 2017). But still, even a healthy human body cannot process the intricacies of smell even remotely as well as, say, dogs can (e.g. Yong 2023: 17–26). Due to their heightened sense of smell, the western Siberian Khanty, for example, saw dogs as interlocutors and guardians between the human and the spirit world (Siikala et al. 2006: 53–54). Some of the same elements are reflected in modern canine police culture, where search dogs allow the police to access the hidden scent-world (Ahto-Hakonen and Hakonen 2022).

Fresh air does not really smell of anything, and yet it is a distinct fragrance, a brisk sensation, which is the first thing usually noticed by visitors to Lapland. Unsurprisingly, Lapland is frequently marketed as having the freshest air in the world. But what is fresh air except a lack of pollutants? Even in the fjells and mountains, some scents, especially of nearby vegetation, are carried along by the winds, but these may not be noticeable enough for the human senses. In our experience of Ritničohkka, the scarcity of vegetation in relation to the sheer openness of the scenery made scents rather homoeopathic, diluting into nothingness.

Before the Russian invasion of Georgia in 2008 and the ongoing invasion of Ukraine, Oula used to frequently work with his Russian colleagues in various parts of Russia—for instance, in the New Siberian Islands of the High Arctic region. One of the first things he noticed was the almost complete lack of smells on the tundra. Early in the Arctic summer, in late June, the permafrost and thick snowbanks emit hardly any scents. As the summer proceeds and the topmost soil and small meltwater ponds appear over the frozen ground, some fragrance of muddy soil and undergrowth emerges. Experiencing this environment for an extended period of time, typically 1.5–2 months at a time (due to the expensiveness of travel to such extremely remote fieldwork locations), Oula's senses grew accustomed to the prevailing diminutive scentscape. This attunement was always disrupted by the return trip to mainland Siberia, involving a six-hour flight in an old Soviet-era Mi-8 helicopter. The related mechanical potpourri of smells of metal, oil, exhaust fumes and flight fuel was quick to override the scentscape of the High Arctic, underlining its fragility.

The motion of air makes scents elusive. In still air, fragrance accumulates as invisible stimulating clouds of vapour. When air moves, these accumulations become less dense, even as sources of smells emit more fragrance as they sway. Indoors, say in a cabin with its mouldy walls, scents are accentuated by the relative scarcity of fresh outdoor scents. Thus, on top of Ritničohkka, the wind strongly shaped our sense of place. In clear weather, the wind was a mild breeze, providing a pleasant sense of comfort. But once the clouds had descended upon us, the wind turned into a buffeting tempest, exuding hostility. Even opening the cabin door became difficult, as the hinge opened to the wind. As you pushed the door open by using your body as a lever, you could feel incorporeal beings— with aggressive gusts of wind—trying to slam the door shut. Outdoors in the gale, you were constantly being grabbed and pulled as if by a physical assailant. The gale was at its strongest between the cabin and the adjacent shed, right outside the cabin door, where a wind-tunnel formed. If such winds had prevailed throughout the mountain, field work would have involved an even greater risk of injury.

The air inside grey rain clouds, such as those covering Ritničohkka, is invariably cold and wet. The water vapour whirling onto a body caught in the middle of a cloud pools quickly on the clothes, conducting the chill of the wind through them, onto skin and tissue. To some extent, moving uses energy and maintains body heat, and, inside a dry cloud, body heat and wind are enough to keep a person from getting excessively wet. But in wet conditions, the balance tips: moisture gathers increasingly while body heat evaporates quicker, eventually forcing the chill to the bones, so to speak, especially if one is not properly clothed.

The chill is a physical experience, which gives rise to mental imagery. Such a state dulls outer body perception, forcing increased attention on the feeling of the chill itself. Attentiveness and patience diminish during the chill, and the sense of time alters, as every uncomfortable moment drags on longer. If a pleasant shelter is not standing by or within reach, the chill is a strong psychological factor affecting how an otherwise stable environment is perceived. Thus, wind and rain are elements whose actions alter the sense of the landscape itself. Through embodied perception, even the incorporeal is observed, laden with agency.

6.7 An Escape Through a Deceiving Land

The next morning, we woke up to continued rainy weather in our thick shroud of a misty cloud. Since there was no way the helicopter could retrieve us from the obscured summit, we decided to "evacuate" the summit and reach a lower altitude. The Lake Háldijávri public wilderness hut some 3 kilometres away was the nearest shelter with potential access for the helicopter.

As we started our descent down Ritničohkka, the weather was turbulent and grey, but the skies and landscape seemed more animated and alive than the day before. As we traversed down the slope, the environment around us was constantly changing, with tattered clouds and cloudy fog opening and closing up on us in an erratic rhythm. This provided glimpses of the surrounding landscape, which hid seconds later, only to reveal itself again. Every time the landscape revealed itself, it looked dissimilar to how we had just seen it. Messiness and restlessness are probably the best ways to characterize how this land exhibited itself to us. Clouds were coming and going, sometimes descending lower into the valley we were headed towards, sometimes climbing higher, and alternating between showers of rain and occasional rays of the sun let in momentarily by parting clouds.

The first descent, with half of our equipment, took about an hour, though it was difficult to keep track of time. Between minding our steps, observing the spectacle of the morphing landscape around us and hauling the bags—including Roger's flight trolley bag, which was never meant to witness such conditions—it is a wonder that it didn't take longer. Much to our pleasure, we discovered that there was plenty of room in the public hut by Lake Háldijávri, one of many similar open wilderness huts in Finland. The Háldijávri hut could comfortably house some 20 people, which was a welcome change to our previous tight accommodation. Like all the other public wilderness huts, the Háldijávri hut lacks modern amenities—there is no electricity, no running water, even worse phone and internet connection than on Ritničohkka—but it was dry and warm. Remote as the location is, it felt safe and calm. It was, indeed, as if, during our descent, we had crossed a fluctuating border between the isolated and existentially disturbing realm of Ritničohkka and a more stable and familiar human world. Leaving Roger with his trolley bag to guard our luggage, we went back up to Ritničohkka to retrieve the rest of our gear.

As we were coming down the second time, the clouds suddenly parted and the landscape looked different yet again from what it had been only three hours ago on our first descent. The topography was barely recognizable, as it flattened and stretched curiously. It was difficult to get a hold of any familiar forms and shapes on the land. We were tracking our walk on a GPS navigator and could therefore confirm that, according to the Euclidean geometric coordinate system, we were passing the same waypoints we had gone through before, but the sense of the land had shifted yet again. Our spatial comprehension was being distorted in the constantly changing light and cloud conditions. At one point, for instance, Lake Háldijávri appeared to loom very far on the horizon, an estimated 15 kilometres away from where we were standing, although we knew that the distance was only 1.5 kilometres. The skies and weather produced illusions and mirage-like visions where it was difficult to tell what was real and what was not. Moreover, we kept hearing an occasional distant hum that sounded like a helicopter. But surely a helicopter would not be flying in these foggy, overcast and poor-visibility conditions? And, if the sounds were real, we could not tell whether the noise was coming from somewhere nearby or far away. All this made it easy to relate to how travellers in different times have described their encounters with Lapland, musing on how these northern lands were outside the ordinary world, as if in another space and time (Naum 2016).

Our experience of Ritničohkka challenged the sense of coherent reality. The landscape seemed to be constantly shifting and changing, which for us became a real and fundamental aspect to this land, not just a misperception of unchanging reality. Rather, Ritničohkka and the lands around it were actually composed of erratically stretched views, reflections and mirages entangled with specific materialities on the ground as well as below and above it. All these visions were unfolding around us again as we walked down the slope in the landscape for the second time (Fig. 6.3). Like a layered hall of mirrors, which did not obscure the land and hide reality, Ritničohkka, with its dynamism, proposed to us instead that this is what reality actually is, even if at its root it was a seemingly static sea of stones.

It is well established in theory that landscapes are and should be understood as temporal processes rather than static sceneries, as Ingold (1993) argued in his classic article on the temporality of the landscape. But if change is constant, the question again arises: what should we document, and how, when doing fieldwork? What we perceived and felt on the fjell were many things simultaneously: ambiguity and often undefinable identities of the tangible and intangible constituents of our environment, with the persistent feeling that there was more beneath the surface of things than we could perceptually and cognitively grasp. This resonated closely, as described earlier, with how Algernon Blackwood describes the Danube in *The Willows*, with the riverine landscape not only continuously and subtly changing, but also becoming increasingly entangled with an invisible otherworld that leaks into the real world. The Danube landscape as he describes it feels somehow wrong and permeated with a malign consciousness that is embedded in the experienced landscape, exemplified by strange gong-like sounds emanating from both the land itself and from inside the protagonists.

Fig. 6.3 Subjective and objective merge and twist in a messy landscape

Our perception of Ritničohkka was not malign exactly, but neither was it stable, perceptually or ontologically, with the lands and skies intertwined with our bodily-perceptual-cognitive instruments (Fig. 6.4). It became clear to us how the weather, for instance, is not merely something that happens on a stable and seemingly unchanging ground, but everything merges together so that the skies are an integral constituent of the land and inextricable from it. The second descent especially was a moment of a heightened awareness of how the shapes, lights and colours of all things are intertwined, related, animated and alive, with everything constantly changing, shifting and transforming in varying ways and speeds. Like altered states of consciousness, it was revelatory in its disorientation. We knew where we were within a coordinate system, but we were simultaneously visually and perceptually lost.

As we made it to the Háldijávri hut again, one of Oula's old acquaintances from Kilpisjärvi was unexpectedly standing on the porch. This strange encounter was readily accepted by us as part of the strange dynamics of things in this already weird landscape. It turned out that the acquaintance was taking part in a search and rescue mission. A group of Bulgarian tourists had gotten lost on the Norwegian side of Háldi, having driven up as close as possible to the summit from the north and proceeding on foot to admire the scenic view, which must have been a

Fig. 6.4 Skies are alive

two-hour hike away. That is, until they were abruptly swallowed by the clouds.

The unexpected visitors had not been prepared at all for the climate, and the results could have been disastrous had not two Finnish hikers encountered the disoriented group, called emergency service, and led the soaked and chilled-to-the-bone Bulgarians to the safety of the Háldijávri hut below the cloud, where Roger was observing the unfolding spectacle. It had, after all, been the sound of the helicopter evacuating the first set of tourists that the rest of us had heard in the messiness of our sensory world descending the slope of Ritničohkka.

"Háldi deceives people", Oula's acquaintance observed before going to look after the Bulgarians waiting to be flown back to the village of Kilpisjärvi. Much to our relief, as the hut was getting a bit crowded, the weather improved just enough for the helicopter to come and pick up the rest of the tourist group, leaving us wondering about the unexpected encounter and the foolhardy unpreparedness of the southern tourists, who clearly had little experience of the constantly and swiftly changing conditions of the High North.

As if we had not just been chewed up and spat out by Ritničohkka ourselves.

References

Abbott, B.P. 2016. Observation of Gravitational Waves from a Binary Black Hole Merger. *Physical Review Letters* 116: 061102.

Ahola, M., and K. Lassila. 2022. Mesolithic Shadow Play? Exploring the Performative Attributes of a Zoomorphic Wild Reindeer (Rangifer tarandus) Antler Artefact from Finland. *Time and Mind* 15 (2): 167–185.

Ahto-Hakonen, J., and A. Hakonen. 2022. All Fun and Games? Relationships Between Finnish Police Dogs and Their Handlers. *Society & Animals* 31 (5–6): 726–743.

Artress, L. 1995. *Walking a Sacred Path: Rediscovering the Labyrinth*. New York: Penguin.

Aspelin, J.R. 1877. Jatulintarhat Suomen rantamailla. *Suomen Muinaismuistoyhdistyksen Aikakauskirja* II: 115–164.

Baird, D. 2004. *Thing Knowledge: A Philosophy of Scientific Instruments*. Los Angeles: University of California Press.

Baron-Cohen, S., D. Johnson, J. Asher, S. Wheelwright, S.E. Fisher, P.K. Gregersen, and C. Allison 2013. Is Synaesthesia More Common in Autism? *Molecular Autism* 4 (40). https://doi.org/10.1186/2040-2392-4-40.

Conway, J. 2011. Getting Lost in the Labyrinth: Information and Technology in the Marketplace. *International Journal of Social and Organizational Dynamics in IT* 1 (3): 50–65.

Cytowic, R.E. 2018. *Synesthesia*. Cambridge, MA: The MIT Press.

Debord, G. 1956. Theory of the dérive. *Les Lèvres Nues* 9. Available at: https://www.larevuedesressources.org/theorie-de-la-derive,038.html. Accessed 25 June 2023.

de Certeau, M. 1984. *The Practice of Everyday Life*. Berkeley and Los Angeles: University of California Press.

Dixon, D. 2007. I Hear Dead People: Science, Technology and a Resonant Universe. *Social and Cultural Geography* 8 (5): 719–733.

Doner, J. 2022. The Dynamic Uncertainty of Narrative, Place, and Practice in Spiritual Experience: Clues from the Phenomenology of Walking a Labyrinth. *Archive for the Psychology of Religion* 44 (3): 129–146. https://doi.org/10.1177/00846724221131149.

Doob, P.R. 1990. *The Idea of the Labyrinth from Classical Antiquity Through the Middle Ages*. Ithaca: Cornell University Press.

Dowd, M., and R. Hensey, eds. 2016. *The Archaeology of Darkness*. Oxford: Oxbow Books.

Edensor, T. 2005. The Ghosts of Industrial Ruins: Ordering and Disordering Memory in Excessive Space. *Environment and Planning D: Society and Space* 23 (6): 829–849. https://doi.org/10.1068/d58j.

Epstein, R. 2021. Your Brain Is Not a Computer. It Is a Transducer. *Discover*. https://www.discovermagazine.com/mind/your-brain-is-not-a-computer-it-is-a-transducer.

Graves-Brown, P., and J. Schofield. 2020. Encountering Landscape: Travel as Method. *Landscapes* 20 (1): 61–84.

Greenwood, S. 2009. *The Anthropology of Magic*. Oxford: Berg.

Gros, F. 2023. *A Philosophy of Walking*, Revised and Expanded Second ed. London: Verso.

Halpern, P. 2020. *Synchronicity: The Epic Quest to Understand the Quantum Nature of Cause and Effect*. New York: Basic Books.

Harris, P.A. 2014. Tracing the Cretan Labyrinth: Mythology, Archaeology, Topology, Phenomenology. *KronoScope* 14 (2): 133–149.

Harvey, J. 2007. *Photography and Spirit*. London: Reaktion Books.

Herva, V.-P., and A. Lahelma. 2020. *Northern Archaeology and Cosmology: A Relational View*. Abingdon: Routledge.

Herva, V.-P., O. Seitsonen, T. Paphitis, T. Komu, G. Moshenska, and R. Nurmi. 2024. Spiralling into a Labyrinth of Cultural Fantasies and Extractivism: Treasures, Extraordinary Undergrounds, and the 'Temple of Lemminkäinen' (Sipoo, Finland). In *Connecting with Ambivalent Heritage: Creative*

Uses of Postindustrial Spaces, ed. T. Äikäs and T. Matila, 131–154. London: Bloomsbury.

Hoover, K.C. 2018. Sensory Disruption in Modern Living and the Emergence of Sensory Inequities. *Yale Journal of Biology and Medicine* 91 (1): 53–62.

Hürter, T. 2022. *The Age of Uncertainty: How the Greatest Minds in Physics Changed the Way We See the World*. Translated by D. Shaw. Berlin: Scribe.

Ikäheimo, J. 2011. Jatulintarhojen vuosi. *Muinaistutkija* 2 (2011): 62–63.

Ingold, T. 1993. The Temporality of the Landscape. *World Archaeology* 25 (2): 152–174.

Kern, H. 2000. *Through the Labyrinth: Designs and Meanings Over 5000 Years*. Munich: Prestel.

Li, B., M.L. Kamarck, Q. Peng, F.-L. Lim, A. Keller, M.A.M. Smeets, J.D. Mainland, and S. Wang. 2022. From Musk to Body Odor: Decoding Olfaction Through Genetic Variation. *PLoS Genetics* 18 (2): e1009564. https://doi.org/10.1371/journal.pgen.1009564.

Lockhurst, R. 2017. The Weird: A Dis/Orientation. *Textual Practice* 31 (6): 1041–1061.

Luke, D. 2010. Rock Art or Rorschach: Is There More to Entoptics Than Meets the Eye? *Time and Mind* 3 (1): 9–28. https://doi.org/10.2752/175169710X12549020810371.

Marila, M. 2020. On the Nature of the Aesthetic in (an) Art/Archaeology. In *Ineligible: A Disruption of Artefacts and Artistic Practice*, 104–119. Santa Tirso: Museu Municipal Abade Pedrosa.

McCullough, D.W. 2005. *The Unending Mystery: A Journey Through Labyrinths and Mazes*. New York: Anchor Books.

McGann, J.P. 2017. Poor Human Olfaction is a 19th-Century Myth. *Science* 356 (6338). https://doi.org/10.1126/science.aam7263.

Merleau-Ponty, M. 1962. *Phenomenology of Perception*. London: Routledge and Kegan Paul.

Morris, C., and A. Peatfield. 2006. Experiencing Ritual: Shamanic Elements in Minoan Religion. In *Celebrations: Anthropological and Archaeological Approaches to Ancient Greek Ritual*, ed. M. Wedde, 35–59. Athens: Astromeditions.

Morris, C., and A. Peatfield. 2012. Dynamic Spirituality on Minoan Peak Sanctuaries. In *The Archaeology of Spiritualities*, ed. K. Rountree, C. Morris, and A. Peatfield, 227–245. New York: Springer.

Morrison, M. 2007. *Modern Alchemy: Occultism and the Emergence of Atomic Theory*. Oxford: Oxford University Press.

Naum, M. 2016. Between Utopia and Dystopia: Colonial Ambivalence and Early Modern Perception of Sápmi. *Itinerario* 40 (3): 489–521.

Niukkanen, M. 2009. *Historiallisen ajan kiinteät muinaisjäännökset, tunnistaminen ja suojelu*. Helsinki: Museovirasto.

Olsen, B. 1991. Material Metaphors and Historical Practice: A Structural Analysis of Stone Labyrinths in Coastal Finnmark. *Fennoscandia Archaeologica* 8: 51–58.

Papadopoulos, C., and H. Moyes, eds. 2022. *The Oxford Handbook of Light in Archaeology*. Oxford: Oxford University Press.

Paphitis, T., R. Norum, and V.-P. Herva, eds. 2021. *Time and Mind* 14 (3): Special Issue on 'Minding Arctic Fields'.

Pétursdóttir, Þ. 2018. Drift. In *Multispecies Archaeology*, ed. S. Pilaar Birch, 85–101. Abingdon: Routledge.

Schmidt di Friedberg, M. 2017. *Geographies of Disorientation*. Abingdon: Routledge.

Seitsonen, O., and S. Viljanmaa. 2021. Landscapes of Sámi Reindeer Domestication and Pastoralism in the Gilbbesjávri Region, Sápmi, Northernmost Europe ca. 700–1800 A.D. *Journal of Field Archaeology* 46 (3): 172–191.

Siikala, A.-L., V. Napolskikh, and M. Hoppál, eds. 2006. *Khanty Mythology*. Helsinki: Finnish Literature Society.

Tilley, C. 1999. *Metaphor and Material Culture*. Oxford: Blackwell.

Uino, P. 2003. Itäisen Suomenlahden jatulintarhat. In *Viipurin läänin historia 1: Karjalan synty*, ed. M. Saarnisto. Jyväskylä: Gummerus Kirjapaino.

Westerdahl, C. 1991. Lotsning och labyrint. *Ångermanland/Medelpad* 1990-91: 77–95.

Westerdahl, C. 2014. Spiritscapes of the North: Traces of the Fear of the Drowned in Maritime Landscapes? In *Med hjärta och hjärna: En vänbok till professor Elisabeth Arwill-Nordbladh*, ed. H. Alexandersson, A. Andreeff, and A. Bünz, 483–503. Göteborg: Götegorgs Universitet.

Yong, E. 2023. *An Immense World: How Animal Senses Reveal the Hidden Realms Around Us*. Dublin: Penguin Random House.

Open Access This chapter is licensed under the terms of the Creative Commons Attribution 4.0 International License (http://creativecommons.org/licenses/by/4.0/), which permits use, sharing, adaptation, distribution and reproduction in any medium or format, as long as you give appropriate credit to the original author(s) and the source, provide a link to the Creative Commons license and indicate if changes were made.

The images or other third party material in this chapter are included in the chapter's Creative Commons license, unless indicated otherwise in a credit line to the material. If material is not included in the chapter's Creative Commons license and your intended use is not permitted by statutory regulation or exceeds the permitted use, you will need to obtain permission directly from the copyright holder.

CHAPTER 7

Gear Shift: Hiking and Being in the North

The wilderness hut to which we had relocated stands in a sheltered valley only 3.5 kilometres away from where we had been, below Ritničohkka and Háldi, on the shore of the desolate Lake Háldijávri. While the distance between the Ritničohkka cabin and the Háldijávri hut seems negligible, relocating had felt like a transition between two worlds, from the weird and otherworldly reality of Ritničohkka to the more familiar Háldijávri. Walking twice through the boiling mirage-like landscape on our descent from Ritničohkka emphasized this sense of transition, like venturing across a liminal zone or through a portal, which entailed an almost ceremonial walk of a total of 10 kilometres across difficult terrain.

The wilderness hut by Lake Háldijávri and its surroundings felt calm and tranquil. Indeed, the place felt somehow *existentially and metaphysically* safer than the vicinity of the dismembered corpse of the perennial snowfield, with its fungi-spewing cabin and closed-off shed. Grey mist continued to surround us, but the wilderness hut felt like a human world in contrast (Fig. 7.1). However, an afterglow of the weirdness that we had been exposed to was still very much there with us and continued to mediate the ways in which we perceived and related to our new setting. This was natural and expected in many ways: we were still in the same high northern world, despite the figurative "dimensional shift" that had taken place during our descent from Ritničohkka. It was in the Háldijávri hut that we first started processing our experiences on Ritničohkka, looking through the lens of weirding (as discussed in Chapter 2).

Fig. 7.1 Lake Háldijávri wilderness hut

The overall sense of weirdness was nonetheless toned down as we settled in the wilderness hut, which, even though the lost Bulgarians had left, we still shared with a number of hikers, some of whom sought shelter due to the difficult weather conditions. Thus, our attention shifted more towards the social relations between people in the context of "wilderness", on the one hand, and the vast and complex networks and entanglements of people, things, infrastructures and the environment, on the other, particularly as reflected in hiker culture and the ideas and practices of tourism and exploration in a remote High North.

In this last chapter, we settle back into the more or less ordinary world—which does not lack weird aspects—and set our "expedition" in a broader context. We consider how tourists, hikers and travellers to the Upper End "walk" the region, both in a literal sense and in terms of how they perceive and engage with this northern land.

7.1 Public Wilderness Huts in Finland

The Háldijávri hut is part of a public, state-maintained network of wilderness huts in Finland. The Finnish word for wilderness hut is *autiotupa*, which means, literally, a desolate or uninhabited dwelling. These open-access log houses are located along hiking trails and have some basic amenities, usually a fireplace, a table, benches and unfurnished wooden bunk beds, with an adjacent firewood shed and an outhouse. This infrastructure is proudly touted as being unlike anywhere else in the world. In principle, it would be possible to hike from the southern tip of the country to the northernmost point staying in these huts every night. The history of this network of open cabins goes back for centuries (e.g. Nenonen 1999: 360), when they were first built along certain trails, such as the postal and trade routes traversing the roadless Finnish forests and fjells of the north, to provide shelter for any travellers in need.

Today, the network of wilderness huts primarily serves recreational hikers and is maintained by Metsähallitus, the Finnish National Board of Forestry, which replenishes the firewood and empties the outhouses. A specific, albeit largely unwritten "wilderness etiquette" has emerged over the years. For instance, users of wilderness huts must bring their own foodstuff, sleeping equipment and other personal supplies, and they are also responsible for the cleanliness and maintenance of the huts. Visitors are expected to chop firewood for the next arrivals, prepare the kindling inside the fireplace, sweep the floor and take their trash with them upon departure.

This etiquette revolves around respect. Visitors should respect all other users of the hut and make room for new arrivals, especially when the weather is poor. This is also the reason for preparing the fireplace for the next users, so they can warm up quickly in case of an emergency. Public wilderness huts are founded on the broader concept of "everyman's rights", which allows everyone to access and use nature freely, as long as they do not harm it or disturb others. The principle of everyman's rights is formalized in Finnish legislation, but, in practice, it is based on an ethos of reciprocal mutual trust and respect. This reciprocity can be traced back to being not only between people but also between humans and non-humans, an integral element of the northern relational understanding of the world. In more concrete terms, self-maintenance of the huts by a constant, even if infrequent, stream of visitors is crucial to the longevity of the infrastructure.

In addition to unlocked public huts, there are a plethora of other types of cabins in wilderness areas, from privately owned summer cottages to rental cabins. There are cabins all over Lapland, used by reindeer herder cooperatives, tourism operators, and the border guard and other authorities. Telephone companies also used to have maintenance cabins, such as the one on Ritničohkka. These latter cabins have become largely obsolete and are only occasionally used by border guards, who routinely check the condition of the radio tower system. In more than one way, different wilderness huts are much more than just huts. They echo aspects of Finnish and Sámi history and identity as well as the state control of remote areas and critical infrastructure.

7.2 Hallucinatory and Spectral Huts

Wilderness huts and cabins comprise a peculiar sensory and experiential environment, even when they lack the sort of disturbing and monstrous qualities that we encountered in the Ritničohkka cabin. There are usually no fixed electric lights, so the fireplace, candles, battery-powered torches and daylight coming through the windows comprise the sources of light. Because windows must endure polarized elements—freezing cold in the winter and scorching summertime heat, as well as wind speeds ranging from breezes to storms—windows are usually small and paned. These types of windows channel daylight into projected cones on the floor and the opposite wall, which themselves emit light and generate shifting shadows in the hut. Fire, usually kept in a metal stove, dances unpredictably. This movement of light and shadow transfixes the eye and can be hypnotic. Small wonder, then, that hearths and fireplaces are often described as televisions before televisions. The dance of the fire is a peculiar genre of entertainment. Superfluous in its purpose, lacking narrative, it is nevertheless often intoxicating to observe. For safety reasons, the fires burned in the stoves are typically closed behind a small hatch. But the air inlet, kept open to keep the fire going smoothly and to prevent the formation of carbon monoxide, emits a diminished glow of flame-coloured light through the opening, pulsing with yellow-orange-red hues. Some stoves also have small window panes in their hatch, to allow more light into the cabin.

Together, these transmuted shadows and shafts of light make cabins places of fixed and moving shadows. Spectral window muntins and mullions superimpose back-lit crosses on surfaces, and dark corners become ever dimmer outside the shafts of light. Moving one's body over a fixed shadow causes unaccounted wavy movement, sometimes captured and interpreted in the corner of an eye. Light and its movement is translated by the sense of vision into an altered form.

Wooden surfaces, dead things carrying the memory of living organisms, contain altering shapes and shades of colour, formed during the tree's life. These forms interact with light and shadow, merging without actual intent into continuities of shapes and forms. A beam of light streaming through a window sets a glowing spotlight on the floor, with an almost imperceptible planetary motion revealed in relation to the dots and shapes on the wood. Elastic moments of time and movement reveal themselves through the interplay of light and shadow. Motion is accentuated by the rarely constant physique of clouds altering the sunlight emanating through the windows. With the chaotic dance of fire reflecting to the eye from the elaborately decorated veins of carved-up dead organisms, the cabin interior pulses with stimuli. The effect that this has is well encapsulated by the saying "walls closing in on you" or, indeed, "cabin fever".

It makes sense on such experiential grounds that wilderness huts are often described as spooky places. Visitors who come alone to a vacant hut regularly write about their uncanny feelings in the visitor log book. There is indeed a richness of stories of supernatural encounters associated with isolated huts, today and in the past. As it happens, one night Markus woke up in the middle of the night in the Háldijávri hut and saw a dreamy figment of a human-like appearance in the doorway. As quickly as he got his eyes properly open and shook the sleep away, the shady figure was gone. This fleeing spectral presence was likely a product of an altered state of consciousness that, this time, resulted from the in-betweenness of being half-asleep and half-awake.

Alleged ghostly experiences, ghost stories and supernatural experiences are not uncommon in Lapland. This echoes the long-standing perceptions of Lapland as a magical and mystical place on the one hand, and the emic northern view that people co-inhabit the world with diverse spiritual beings and more or less person-like non-human beings on the other. As for weirding, ghosts are reminders of how time is not simply a linear arrow; rather, past and present are intertwined in many ways, and the former has a presence in the latter, sometimes manifested in ghostly or spectral forms. Human company often mutes such ephemeral, strange phenomena or experiences. But wilderness cabins are unusually hallucinatory places: beacons of safety, often off the grid and with lingering elements of uncertainty and uninvited presences just beyond the range of vision. In this way, wilderness huts sustain traditions of supernatural stories and extraordinary feelings and experiences that date back centuries, to when people huddled in their dark and smoky log houses.

7.3 Human Encounters

The Háldijávri hut comprised two parts: the open wilderness hut side and the rental hut side, where one can book a bed and acquire a key in advance for a small fee. When we arrived, there were a few hikers in the rental quarters, and a handful of people joined us in the open hut side after we settled there. The weather conditions prevented us from conducting further surveys and discouraged most of our new acquaintances from continuing their journeys. There were rumours of a swelling river blocking the route back south. So we felt stranded again—only stranded with other people this time, whether we liked it or not. Based on their stories, most of our companions seemed to feel more or less unsure in the circumstances, having been pulled to this remote nook of land by some personal quest, which the weather conditions now hindered.

Wilderness huts are both social and asocial places. People come from all around the country—sometimes around the world—and there is little chance of randomly encountering someone you know, except for those you might have just met on the trail. Consequently, most encounters and interactions are between complete strangers. Here, the aforementioned etiquette, the set of unwritten hikers' rules, comes into play again. The main rule is that the first to arrive should always make room for the next—"first come first leave" when a hut reaches its maximum capacity of visitors.

This constant play of musical chairs can sometimes lead to conflicts, which, according to a recent report by the Finnish public broadcasting company YLE, have been on the rise in recent years, as hiking became increasingly popular in the wake of the Covid-19 pandemic and the associated travel restrictions to other countries. Beyond that, the number of visitors to national parks and the most popular hiking trails in Finland has risen annually throughout the 2000s, with fewer people overall familiar with the nuances of the unwritten code. Thus, nowadays, people travelling in groups or pairs often look forward to empty huts and wish to minimize dealing with strangers and sharing space with them. On the other hand, these unexpected encounters provide a chance to exchange knowledge about the trail ahead, tell stories and give hints, or trade excess matchbooks and provisions—or, in our case, make ethnographic observations about our fellow travellers (less formally known as chit-chat). Sometimes temporary friendships are even formed during these fleeting encounters, and hikers often wait at the next cabin to see that others they met on the trail actually make it to their next destination too.

Our small community in the Háldijávri cabin formed for three days, courtesy of the hostile weather conditions, which are not uncommon in this area. The hikers were anxious to continue their respective journeys. But we were not going anywhere, as we waited for either better survey conditions or the helicopter to come and pick us up on the fifth day. Under the circumstances, weather emerged as a natural topic of discussion among the temporary co-inhabitants of this compact world inside the hut. The fireplace attracted various clothes, hats, socks, pants and hiking boots that needed drying. The dangling clothing and accessories inspired talk about how different people had prepared for their hikes, which in turn afforded some insights into the hikers' perceptions of and relations with this land.

The main crowd that stayed in the hut, besides ourselves, comprised a middle-aged mother with her 20-something daughter, two women who were friends and probably in their early 30s, and a young woman hiking alone. Based on the casual conversations we had with our co-inhabitants, a romanticized image of Kilpisjärvi and Háldi (see Valtonen 2019) had played an important role in attracting them here. For most of them, this was their first time in the region, although everyone had evidently spent a significant amount of time planning and preparing by consulting maps, blogs and books, as well as spending a substantial amount of money to gear up with modern hi-tech hiking clothes and equipment.

Our own perceptions of this region were, of course, also mediated by our own imageries and imaginaries, as detailed earlier in this book. What we had going for us was, first, that some of us had been exposed to similar mountain environments before, and, second, that we were relying on a helicopter as our main mode of transportation. So the challenges we faced were quite different from what hikers faced. For them, the reality of the region and its unforgivingly wet and windy open tundra can only be grasped by facing the worst the trail has to offer. Nonetheless, there are ways to potentially circumvent these elements through the promise of modern technology.

7.4 Survival Tech

Clothes and other related gear are elementary to establishing a hiker identity and "hiking credibility". They are also a medium for encountering the world and the experience of it. Hi-tech outdoor clothing serves not only a practical purpose, but also symbolically transforms people into travellers—and "conquerors"—of the "otherworldly" and unfamiliar wilderness—even as, in some sense, it isolates them from it, similarly as a space suit would do. These special clothes foster the illusion of a bounded, self-contained existence, which contradicts the ontological openness of the relationally constituted northern world. In difficult weather conditions, however, even the hi-tech clothing fails, with soaked clothes and boots revealing a rather more disturbing and powerful reality. Suddenly, the senses are triggered by much more than the beautiful scenery to be admired from the illusory safety of the hi-tech gear. As rain seeps through the protective layers of clothing, people become connected and one with the landscape in a new and very concrete way. In one's ordinary, everyday environment, getting wet might be a nuisance or physically uncomfortable. Yet, in the more extreme conditions of a remote northern land, the same experience comes to have an existentialist dimension of horror, which, in addition to purely practical considerations, probably partly explains the pronounced attention to clothes in the wilderness hut.

The North is a polar world. To travel in the North is to travel by season and, most importantly in the short term, within the parameters dictated by the weather. Weather may change abruptly, making gear that was appropriate and sufficient an hour ago utterly obsolete. Moreover, summer and winter hiking gears are specialized and very different from each other. Hiking boots have become somewhat of a conflict zone in this regard. Many hikers rely on technical equipment, including Gore-Tex hiking boots. This reliance clashes with the strictly ecological view of many hikers, who strive to gain a connection with nature through bodily experience. Gore-Tex is coated with durable water repellent (DWR), a PFAS fluoropolymer, which has turned out to be highly toxic (e.g. Lohmann et al. 2020; Lohmann and Letcher 2023). Such toxins gradually leach from the equipment—both boots and other clothes treated to make them hydrophobic or water resistant—possibly even ending up in the environment. As a result, some choose to treat their hiking gear with other more environmentally safe substances, such as processed beeswax. Such methods may not be as effective and durable, but they provide vastly better water resistance to hiking boots compared with regular untreated leather. And, on the plus side, soaked boots are said to dry faster if they have not been coated with DWR.

Keeping oneself from getting soaked is one of the most common practical objectives in hiking in Lapland, and hikers take many precautions to shield themselves from bad weather. Again, the most technical-minded hikers use Gore-Tex-treated jackets and hiking pants. These tend to be more expensive than regular hiking clothes. And for pouring rain, Gore-Tex is usually not enough, as rainwater pools into creases and finds its way through worn-out fissures in the chemicals and cloth. If clothes do get wet, even if a hiker's boots manage to resist absorption, water will trickle down their leg into the hiking boot as well. Especially in cold weather, this is rather catastrophic for many hikers and tends to momentarily turn the hike from a meditative yet arduous delight to a sheer trial of mental and physical survival.

Thus, proper rain gear is essential in the North. In the past, rain clothes used to be made almost solely from rubber. The material was always rather uncomfortable for many reasons, not least because wearing rubber rain gear makes physical activity in the rain a zero-sum game. On the one hand, the rubber protects you from the rain, but it also prevents your body heat and sweat from evaporating, making you hot and wet. On the other hand, if you stop to wait out the rain, you will quickly notice how the rubber conducts the chilling temperature of the rain and wind. So rain in cold weather becomes a hazard for those outside their comfort zones: continue walking and you may need to take off clothes underneath the rubber rain gear, but stay put and you may have to put more on, all the while managing to not get wet.

The development of modern weather gear for hikers has resolved some of these issues. These clothes are breathable, allowing body heat and sweat to pass outside, while preventing water and wind from going inside. This type of technical clothing, much thinner than rubber clothes, also takes up less space. Most hikers, as well as field workers, consider technical breathable weather gear essential equipment.

But hidden beneath breathable hi-tech wear, in the fabric and the chemical coating, is a vast network of interdependencies. Research laboratories, chemical factories, assembly lines and ocean-crossing freight ships are carried into the field as an overarching network that clings to the wearer like a spiderweb. This network though, unlike a spiderweb, can very easily go unnoticed.

The question remains whether we, as fielders, can truly empathize with past people traversing the northern landscapes, mentally attuning ourselves to their experienced circumstances, when we are in fact wearing a second skin of an interconnected global material network that is designed to distance the wearer from the elements. Through our material choices, we realized, we had weirded our own body perception long before our arrival at Ritničohkka.

7.5 Hauling a Mobile Home

The issue of space and weight in hiking is always acute. You are carrying everything you need to sustain yourself on your back. In essence, the body becomes a mobile home.

This was only partly true for our team. As we arrived on Ritničohkka in a helicopter and were supposed to leave the same way, we had taken much more stuff with us than is usual for a hiker (Fig. 7.2). Indeed, any sensible hiker does not carry luxury consumables such as a box of wine with them. As a result, as described earlier, we had to make two trips between Ritničohkka and Háldijávri to relocate all our gear, which, in an ironic twist of our being prepared for almost anything, we had not been prepared to carry on our persons. Roger's rolling suitcase was very much a case in point. It was such an (otherworldly) comical sight: our team descending from the grey wet clouds with Roger dragging along the wheeled suitcase, as if he had been looking for his airport gate and gotten spectacularly lost on the way.

To transform one's body into a mobile home, a proper carrying capacity has to be established. The relationship between the weight of the gear and the space it takes up can always be improved by "gearing up". Other factors in the equation are the cost and effectiveness of the individual articles of equipment. One joy of hiking, especially off-season, is preparing for the eventual journey. Improving carrying capacity for the next hike is an important part of developing as a hiker. While this improvement can also involve physical training, which helps to carry heavier loads, equipment that can be squeezed into a tighter space is constantly being introduced to the market, making the matter of space a technological issue.

All the equipment has to fit into a backpack. The backpacks favoured by hikers today are mostly internal frame backpacks, a leaner alternative to the old-fashioned external frame backpacks. Backpacks tend to

7 GEAR SHIFT: HIKING AND BEING IN THE NORTH   163

Fig. 7.2 Our five-person team may have had more luggage with us than your average hiker. Various backpacks, duffle bags, bundles and rucksacks of assorted fieldwork gear, with Roger's green rolling suitcase in the middle, unloaded from the helicopter on the summit of Ritničohkka

become highly personal. Through repeated use, a hiker becomes intimately familiar with the infrastructure that their backpack offers. Every pocket and every nook will eventually come to use, and every fault will reveal itself. The frame, whether internal or external, distributes the weight of the backpack evenly, relieving stress from the shoulders to the upper and lower back and the waist. Thus, a good framed backpack is made to become part of the person carrying it.

The technology itself is ancient. The famous ice-bound mummy known as Ötzi, discovered in the High Alps, was found with what was apparently an external frame backpack, with the only preserved part being the wooden frame itself. There is thus an unexpected continuity through millennia, despite the vast chasm between how the material networks operate in the two temporal contexts. Hiking as an activity—using one's body as a vehicle to transport not just oneself but also a whole inventory of artefacts—clearly retains some elements that can bridge the gap between the present and the far past. Crossing this bridge can accentuate the feeling of disorientation that arises from being aware of the material entanglements inherent in hi-tech equipment.

The weight-space-cost dilemma is particularly acute regarding camping equipment, especially tents and sleeping bags. The principal is: the more you are willing to pay, the smaller and lighter the equipment. Effectiveness is also an issue. Tunnel-tents—common for hikers, but also used by part of the team on Ritničohkka—are better suited to withstand winds than most other tent types. Windy conditions are inescapable in mountainous regions. But even the most expensive and advanced tents fail eventually when exposed to the elements. One of the reasons we decided to leave Ritničohkka for the more protected Háldijávri cabin was out of worry that the howling, tempestuous winds inside the cloud would eventually tear the fabric of the tents. One of the tents eventually did collapse, but this happened at the Háldijávri hut when the rescue helicopter carrying the Bulgarians passed overhead.

With all this equipment scaffolding a proper hike, the technological aspects of the activity become apparent. While hiking is often presented as the means to escape modern society and technology, however briefly, it is inseparable from technological engagement. Modern hiking gear is an industrial product, researched and developed to the cutting edge. Moreover, it is marketed and distributed in the Western economic framework by private companies engaged in increasing their profit margins. As such, hiking as an activity has two meanings: the more apparent is a return to nature, while the other, hidden in plain sight, is the extension of the prevailing economic system. In fact, different brand names and logos are carried into remote places by both hikers and researchers doing field work.

Indeed, walking in a seemingly remote landscape, our team brought along with it the material-economic networks that expand across the globe, which radiated off our hybrid physical manifestation onto the landscape itself—or, at the very least, our perception of it. The weirdness we experienced in terms of the landscape affecting us: was it merely the landscape reflecting our own weirdness back onto ourselves?

Still, a perhaps even weirder aspect that our stay at the Háldijávri hut made us think about deeply was our changing relationship to the very activity of walking.

7.6 Cultures of Walking

The feet do most of the work in hiking, at least in the most obvious sense. Walking has arguably become an issue of heritage over the last century, as motorized transportation gains more and more ground throughout the world. The availability of cars, bikes and public transport has threatened to make long-distance walking obsolete as a mode of transportation in most of the Western world. In fact, this is regarded as a reason for serious health concerns, not only in Finland and the Nordic countries but globally. People who don't have to walk usually don't. At least not initially. It was only after a public health crisis began to emerge in the West, with effects including diminishing physique, increased obesity and weakening bone resilience (see, e.g., WHO 2022; Burhaein et al. 2024), that walking has come back into fashion.

Walking is often described as a form of meditation. For Doner (2022: 141), "Walking engages the whole person. It transforms and organises experience, stimulates mind, memory, and feeling. We walk, we perceive. We walk, we think. We walk, we feel." Apart from physical activity, walking is a particular mode of perceiving the surrounding world and oneself in relation to it (Vergunst and Ingold 2008). Due to its repetitiveness, walking causes (or enables) the mind to wander. With a consistent change of scenery, the walker gets constantly bombarded with multisensory outside stimuli. The stimuli comes from the landscape, the weather and other agents in the immediate surroundings, while the walker also affects some of these agents in return. The activity itself constitutes an energetic engagement with the world outside oneself. But as a physical activity, walking is also a form of engaging with oneself in more ways than one.

In the summer of 1805, in the waning days of walking as a mode of long-distance transportation, a German traveller and writer Johann Gottfried Seume embarked on a tour of the Baltic Sea. He started from Leipzig, proceeded through the Baltic to Saint Petersburg, went around the southern coast of Finland and arrived by ship in Stockholm, from where he travelled back south (Seume 1806). Despite having previously conducted a nine-month walking tour to Sicily in 1801, he professed to travelling more by carriage on this trip, though he still managed an impressive 1100 kilometres on foot (Valli 2018). Although he did not travel farther north than the southern coast of Finland, his writings still convey the same meditative aspects as hikers to Lapland experience today.

Half a century later, in a lecture-turned-essay, the American writer Henry David Thoreau pondered on walking as a vanishing mode of transportation. In the essay, *Walking*, Thoreau (1862) introduced the meditative aspect of walking to the wider reading audience of his time. He saw walking as a way to connect with nature and escape the strict confines of society. His work formed much of the theoretical basis for modern hiking, along with the idea of hiking as a means to temporarily disconnect from common society.

In Sweden and Finland, most summertime land travel that did not involve hauling cargo was conducted on foot until the nineteenth century, even though footwear was a rare luxury for the average peasant (e.g. Nenonen 1999: 291–292). Lack of shoes has been a common theme in many historical situations. One of the deciding battles of the American Civil War is said to have been initiated by the need for shoes. According to the legend, the Confederate Army, lacking proper footwear and having rapidly walked over 400 kilometres, from Richmond via the valley behind the Blue Ridge Mountains to a place 100 kilometres northwest of Washington, D.C., proceeded to the small town of Gettysburg because it was said to have a shoe factory (e.g. Keegan 2009: 191). Whether this is actually true is beside the point. The more pertinent historical fact is that war and the movement of great cohesive armies were experienced on the ground by individual soldiers as seemingly endless walking. Not everyone could persevere, with "straggling" becoming a major issue for both armies. Stragglers were usually soldiers who became too exhausted, or their feet too sore, to continue. Managing to keep up was brutal work. There are reports of largely shoeless Confederate armies marching dozens of kilometres even during wintertime, leaving a trail of blood from bare broken feet in their wake (e.g. Gwynne 2014: 492). This was and still is a peculiar point of pride for U.S. Confederate sympathizers (e.g. Wiley 2004 [1943]: 81–82, 88–89, 120–122).

Finland, too, has its own heroic lack-of-footwear stories, especially from the Second World War. Footwraps (which have been eclipsed by socks today) are often brought up as a sign of the common-sense ingenuity and resilience of the Finnish soldiers fighting the Red Army and, subsequently, the German Wehrmacht. Footwraps were durable, warm yet breathable, easy to wash and dry, and customizable to the foot. They are often presented in public discourse as an example of how Finland "beat" the far larger Soviet Union in a war despite Finland's meagre resources, thanks to better skills and knowledge of how to survive in the wild. In the end, according to the weird logic of patriotic national myth-making, the Soviet Union was beaten so badly that Finland had to surrender to it, twice. It turned out that Soviet soldiers also used footwraps, which are, in fact, known in French as *chaussettes russes*, "Russian socks".

Footwraps are still an efficient and functional piece of clothing for walking in the wilderness. They dry quickly by an open fire and, if you know how to tie them properly around your feet, are more comfortable than socks and do not easily cause blisters. In fact, many northern Finnish hunters continue to use footwraps to the present day, carrying on the age-old tradition. However, stores selling hiking gear offer wide selections of at least seemingly high-quality socks, including expensive hi-tech solutions, such as breathable waterproof socks.

Walking is both a skill and an ability. As a skill it can be perfected by repetition and exposure to difficult terrain. Fieldworkers often become experts in walking, minimizing injury risk the more they practise the skill. But fieldworkers rarely compare with "indigenous" walkers. On many occasions, carrying out communal archaeological field trips, we have been guided by elderly local informants. Most often, they see the forests and fjells as part of their extended homes, as familiar as a backyard. Once, when we were walking through the woods with an elderly guide during a survey, he commented sardonically that "even if some people seem young on the road, they are old in the forest". He had previously served as a guide to a Finnish film crew and a celebrity host, who struggled to keep up with the elderly man in the pathless and uneven forest terrain, even though he was twice as old as the others.

Some aspects of walking are abilities rather than skills we learn. These abilities are something we adjust to. Feet are malleable. The more you walk, the more stress your feet can take. Even better, the more you walk barefoot, the thicker and more durable the skin of your feet becomes. The ability to walk barefoot has been a niche trend for many years now, made popular especially by barefoot running. Barefoot running was introduced to the general public through books such as *Born to Run* (McDougall 2009), which asserted that running barefoot was the mode of transportation for which the human body had most effectively evolved. This becomes clear when observing the Maasai in East Africa, for instance, who prefer to walk barefoot or in flimsy and seemingly uncomfortable car-tire flip-flops over daily distances of some 60–70 kilometres (Fig. 7.3). In this sense, even the relatively hi-tech hikers of the North may be striving

Fig. 7.3 A Tanzanian Maasai friend of Oula, Israel ole Mollel of Engaruka, walking across the savannah in his car-tire flip-flops while assisting in archaeological fieldwork, with Mount Kilimanjaro rising in the distance

to find not only peace and meditation in their travels, but a return to a "pre-Western" mindset of walking as well.

7.7 Artefacts of the Foot

One of the best ways to attune your feet to walking barefoot is to start by stepping shoeless on a pine cone. First in a controlled fashion, slowly, lightly and surely, with the sole of your heel, which is usually the most resistant part of the foot. Initially the pain is intolerable; and, after enough repetition, you can transition to stepping on the pine cone with the soft parts of your feet. Eventually you should start to develop a thicker skin, which allows you to walk barefoot even on gravel. This rather unfamiliar activity can be considered a basic form of body-engineering, blurring the lines between the human body and technology.

To a person attuned to barefoot walking, different terrains offer different challenges and conditions. Take a walk in a coniferous forest and you may be surprised. Walk on the roots of the trees and you'll notice that the resin from the trees sticks to your feet. You can try to avoid the roots, but walking barefoot you will become acutely aware that they are, in fact, everywhere. As you walk, the resin will start to cover more and more of your soles. And, as you go, needles, tiny twigs, flakes of chipped bark and dead plants will stick to your feet, forming a layer of its own. Nature's own protective layer.

You have now become part of the forest, or rather the forest has attached itself to you. As such, you may feel your nature-culture boundary narrowing.

On Ritničohkka, though, like in many parts of Lapland and the vast mountain ranges of Scandinavia, the only protection for your feet would be the ever-present dead-coral-coloured lichen. Lichen is the mainstay of the reindeer diet, and it provides a soothing surface to walk on, especially in dry, warm weather. Yet, in rocky terrain, shoes are still a requisite for most people, even though this may not have always been so. Remember, walking is both an acquired skill and adapted ability, and in the past, advanced bodily experience of both may have given local travellers in their traditional, thin and soleless homemade leather footwear the means to navigate even the sharp-edged boulder fields of Ritničohkka unharmed.

Modern hiking boots, on the other hand, separate the human from the terrain in a profound way. No longer do your feet attune themselves to the terrain, but rather they adjust to the boots. The boots reciprocate, adjusting themselves to your feet in return, becoming a very personalized piece of equipment (Ingold 2004).

Most hiking boots are made primarily of leather. Leather is what gives them both firmness and elasticity. This is because leather used to be the skin of a living being, much like the skin of your feet. As modern manufacturing methods and material networks are convoluted, this living being, the animal that a particular hiking boot is made of, is anonymous and unfamiliar to the user, further anonymized by the use of chemicals to remove the scent of dead skin. Yet the boot still carries on the biography of the animal. Even in the modern Western world, a leather boot, unconsciously or not, embodies the animal's spirit.

Many people of the North have their own traditional footwear. The Sámi have often used reindeer leather or seal skin as footwear materials. The difference between the traditional, locally made boot and a Western hiking boot is that the artisan crafting the traditional boot may even have been involved in killing and skinning the animal. Thus, the spirit of the animal is ever more present. In many contexts, wearing the animal is associated with its power passing momentarily to the human (e.g. Dönmez 2018; Yusha 2024). Hiking boots in a place like Ritničohkka provide the wearer with some of the animal's adaptive abilities.

Wearing good hiking boots makes even such a hostile place accessible without your feet needing to attune themselves to the terrain, which would require a lengthy process of acculturation and acclimatization. Yet, in some circumstances, even hiking boots fail. Modern hiking boots typically have thick, sturdy and rigid rubber soles that isolate and insulate the hiker from the soil, rocks, roots and other irregularities of the ground below. It is most often this protective insulating rubber layer, buffering the wearer from the ground, that fails. When shoe soles crack or become completely detached, it is difficult to fix them in the field, revealing a dependence on technology not unlike the one revealed by a mobile dead zone.

Reliable boots are perhaps the single most important item to hikers, which goes some way towards explaining why footwear featured to such an extent as a subject of discussion in the Háldijávri cabin. Shoes in general are known to be imbued with cultural meanings (Ingold 2004). While we were wondering about the different dimensions of hiking boots in the broader context of shoes and their cultural-cosmological meanings, we turned yet again to ChatGPT for another attempt at a non-human perspective. The AI offered the following:

> For some, wearing hiking boots is a way of expressing their love and respect for the natural world, and acknowledging their dependence on it.
> Furthermore, hiking boots can also be seen as a way of participating in a larger cultural tradition that values exploration, adventure, and personal growth. This tradition has deep roots in human history, from the early nomadic societies who traversed vast distances on foot to the modern outdoor recreation industry that encourages people to explore the wilderness and push their physical and mental limits.
> In this sense, hiking boots can be seen as part of a larger cultural narrative that values the outdoors as a place of transformation and discovery. The act of putting on hiking boots can be seen as a ritualistic act that signifies a willingness to engage with the natural world, and to undertake a journey of self-discovery and personal growth.

Curiously enough, this AI-generated text undoubtedly captures the very essence of hiking boots.

During our survey, we also came across an old lichen-covered shoe related to the post-Second World War reindeer herding activity in the area. This old, unpaired shoe, found next to a temporary shelter built by herders in the post-war years, represents an intriguing example of the scalability of the surrounding barren landscape (Fig. 7.4). Namely, when examined closely through the microscope's eye, the upper surface of this forgotten shoe reveals a barren landscape in itself, with its own miniature lichen terrain with valleys and peaks, mirroring the treeless tundra landscape where it was found. The landscape, as it turns out, had taken over, conquering what remained of the proverbial footprint of the anonymous walker of the animal spirit.

Fig. 7.4 An abandoned shoe entangled with its environment (left), featuring its own lichen cover, which, under the microscope, reveals a micro-landscape that parallels the barren valleys and peaks of the surrounding land (right)

7.8 "HUIPUTUS" AND OTHER TALES OF CONQUEST

In a sense, Arctic exploration has been inextricable from sport for a long time. Originally very much a colonial venture—like archaeological and anthropological fieldwork, in fact—it still holds a sense of conquering and domesticating the wild unknown. As our team embarked on our journey, the underlying attitude, which we had to consciously restrain, was that we were conquerors on a quest to tame the land through practices of knowledge production. In the end, however, the feeling was, rather, that the land conquered and shaped us in many weird and uncanny ways, as recounted in this book.

What we initially set out to do—although this was not how we consciously thought about it at the time—echoes the hiker (or more generally tourist) practice known in Finnish as *huiputus*. This literally means "to deceive" or "to sham", or to fool or cheat, but it is also a word play meaning "topping the peak" (*huippu* means "top" or "peak" in Finnish). *Huiputus* in this sense is effectively a light-hearted version of "conquering" a landmark. The "conquest of Háldi", for instance, is a commonly recognized achievement among hikers. And some of our co-inhabitants in the Háldijávri cabin were indeed on this quest to conquer the mythical Háldi. They themselves did not phrase (or even mean) it quite like that, talking about it more as a special personal and exotic experience in a landscape quite unlike what they could find closer to home. Mountaineering is, traditionally, rife with the language and imagery of conquering, overcoming and taking over, which have clear colonialist connotations. This is troublesome in settings like Sápmi/Lapland, which for centuries have been subject to Nordic colonialist efforts that, even today, are often not even recognized as colonialism by the general public.

The language used by hikers, who typically are not locals, illustrates how there is still a tendency to approach the land and landscape as a commodity waiting to be conquered by a "Great White Hunter". This, in fact, closely mirrors the more general attitudes to Lapland, Sápmi and the North since early modern times (e.g. Naum 2016). These lands have been, and still are, often presented to tourists and other outsiders as a blank slate—empty wilderness waiting to be taken and put to proper use, either as a tourist destination or as a source of unused raw materials. In these kinds of generalizing views, local people—whether Sámi or Finnish—are often almost non-existent, mirroring how Lapland is marketed as "pristine" untouched natural wilderness, devoid of human touch or history.

Group hikes are organized commercially for various experience levels. Tourism entrepreneurs offering these trips typically call them "trekking products". This underlines the commercialized, monetary and capitalist aspects of these wilderness activities. Together with the distant drone, often audible while walking in the fjells, of helicopters and seaplanes ferrying trekkers and fishers, commercial hikes illustrate how global capitalism reaches even the remotest, roadless fringes of the northern world—using even unwitting researchers as its vessel. These tourism activities overlap with more traditional land use practices, most notably the Sámi reindeer herding that takes place in the same general area. For many hikers, encounters with the semi-domesticated reindeer are among the many charms of Lapland.

There are signs of rising cultural awareness among tourism entrepreneurs. One of the tourist guides in Kilpisjärvi, who offers guided tours to Háldi and other destinations, has decided to stop using the term "conquest" in the names of his advertised treks. In 2022, he rebranded his top product from "Conquest of Halti" to "Trek to Halti" (Veijonen 2022). In a blog post, he explained that he originally named the trip "The Conquest" to honour the first Finnish hikers—three women, in fact: Kaarina Kari, Anna Lehtonen and Inkeri Arajärvi from Southern Finland—who trekked there in 1933 with Sámi guide Pietari "Piera" Proksi from Näkkälä village. In 1978, Kaarina Kari published a book about their endeavour titled *Conquest of Halti* (Fi. *Haltin valloitus*).

Kaarina Kari—a pioneer of Finnish hiking culture—is an intriguing case in hiker lore. She never got married, living instead with her friend Anna Lehtonen her whole life, and from their private correspondence it seems very clear that they were a couple, a fact they kept secret from the unapproving society (Ranta 2023). In naming his product, the tour guide wanted to honour the achievement of Kari and the all-female team, even though he himself does not believe in a "conqueror mentality" in trekking. Instead, he explained, "[f]jells are not conquered, they either let us in or they don't. Nothing is taken for granted in the harsh conditions of the great fjell region" (Veijonen 2022). However, as he tells it, in 2022, one of his customers had asked: as the "white culture has conquered the Sámi lands, is it worth repeating the discourse in the name?" The guide could not stop thinking about the remark. Owing to this eye-opening incident, he decided to change the names of all his trekking products to avoid reproducing and contributing to colonialist discourses (Veijonen 2022).

7.9 "Big-Man" Legends of the Trail

Finland's history in the latter half of the twentieth century is in many ways intertwined with the former—and somewhat mythologized—president, Urho Kaleva Kekkonen (1900–1986), and his wide-ranging undertakings. Kekkonen was the longest-serving president in Finnish history, in office from 1956 to 1982—years later remembered as the time of "Kekkoslovakia", in joking reference to Czechoslovakia, which was politically in the sphere of the Soviet Union. His personal relationships with world leaders, especially of the Soviet Union, and a cult of personality developed by the state and media came to define the image of Finland during the Cold War (see Lähdesmäki 2009). Kekkonen was consciously presented as "the mighty leader of his pack" (Suomi 1988: 580), a man of the common people from a poor background. Special emphasis was placed on his apparently mythical and superior sportsmanship, masculinity, mental and physical strength, and especially his survivalist skills (Kuronen and Virtaharju 2015; Wuokko 2011).

This survivalist image was partly built on his annual spring skiing trips to different parts of remote Lapland, frequently to the "Upper End" in the Kilpisjärvi region. On these trips, which began in 1961, Kekkonen traversed landscapes similar to those we surveyed. He "conquered" Háldi and most often used a cabin some 15 kilometres south of Ritničohkka as his basecamp. There is an out-of-place memorial plaque to Kekkonen on top of the adjacent Čáivárri (Saivaara) fjell, which clearly echoes the cult of personality. This monument and its background add a curious layer to this enchanting northern land, using it not only as a material resource—as in the case of mining and tourism—but also as a symbolic one. To exploit these resources, even the state apparatus remains highly invested in the conquest of the High North.

Wherever he went, Kekkonen was followed by an entourage of political, business and military elite. They were known as his "trailing skiers" (Fi. *perässähiihtäjät*), a term which today has strong allusions to sycophancy. Kekkonen was strict about always skiing as the point man, with his followers straggling behind. Taking long ski trips, exhibiting his physical prowess, became a vital part of his conscious myth-building (Wuokko

2011). On these trips, Kekkonen's party regularly occupied public wilderness huts. Despite the common etiquette and despite Kekkonen's image as a man of the people, the huts in which he stayed were reserved exclusively for him and his guests, who often included VIPs like the King of Sweden and the ambassador of the Soviet Union. Accompanying border guards prepared his ski tours weeks in advance, organizing the transport of abundant food and alcohol to the huts ahead of the president and his trailing entourage. Even the ski tracks were well prepared, with a secret vanguard making sure the trail was perfectly smooth. The vanguard was, of course, not allowed to be seen skiing ahead of the supposedly trailblazing strong man (Fig. 7.5).

During his skiing trips to the "Upper End", Kekkonen was given the honorific title of *väärti*, a partner in a reciprocal friendship system of the Sámi that seems to have ancient roots (Länsman 2004). The title was granted by the famous Sámi elder Ántt' Ásllat Gáijohaš (1887–1969; Fi. Aslak Juuso Kaijukka). Gáijohaš was known as the "Patriarch

Fig. 7.5 President Kekkonen (on the right) and his entourage visiting a Sámi tent in Gáijohaš village at Ávžžášjávri on his skiing trip in 1977 (Finnish Heritage Agency HK7966:34/Einar Närhisalo/CC BY 4.0)

of the Upper End". Kekkonen frequently visited Gáijohaš's encampments to socialize with him and offered to have the state build a road to his village in Ávžžášjávri (Fi. Raittijärvi) some 30 kilometres south of Ritničohkka. Gáijohaš declined this offer to prevent outsiders from entering the territory and instead asked Kekkonen for a telephone line, which the state diligently provided. Today, Ávžžášjávri, now inhabited by Gáijohaš's descendants, is the remotest village in Finland and still without a road connection.

The memorial plaque in Čáivárri was originally designed to be a larger-than-life monument, fitting Kekkonen's larger-than-life reputation. It was designed by the world famous Finnish designer Tapio Wirkkala (1915–1985). The monument would have consisted of five boulders, symbolizing Kekkonen's five consecutive terms as president and described as "court stones" (Fi. *käräjäkivet*, a semi-mythical type of archaeological monument in Finland, echoing a Norse *thing* assembly). The path leading up to the stones would have had 10,227 steps, the same number as the days of Kekkonen's presidency. Apparently, Wirkkala had planned to make the monument respectful of and inconspicuous in the local landscape, but the plan met fierce resistance from the wider public. By that point the untouchability of Lapland's wilderness had become common lore among hiking enthusiasts of the south, who were mostly against the monument. Due to the opposition, the plan was never realized, and a small memorial plaque dedicated to Kekkonen was placed on Čáivárri instead (Fig. 7.6).

Another local memorial for Kekkonen is an old abandoned outhouse that has been kept as a relic from his visits (Fig. 7.7). This outhouse is located in the yard of the Kemijoki Ltd. corporation's wilderness cabin, which Kekkonen frequently used as a base on his trips to the Upper End. The outhouse was on the verge of collapsing years ago and a new one was built. But when the builders were about to burn the old one—a typical way of disposing of materials in the wilderness areas—an old border guard, Risto Anunti, stopped them:

> Don't burn it, but tie it to a tree with a chain. There is no other outhouse like it in Finland [or the world], where Kekkonen, Koivisto [the President of Finland after Kekkonen], the King of Sweden, countless government ministers, and Stepanov [a Soviet ambassador to Finland] have visited. (Erkkilä and Iivari 2019)

7 GEAR SHIFT: HIKING AND BEING IN THE NORTH   177

Fig. 7.6   A view of Čáivárri

Fig. 7.7   A view of Háldi (on the far left) and the slope of Ritničohkka (on the far right), with Oula racing to an outhouse which postdates Kekkonen at the Lake Háldijávri hut

7.10 THE SOCIALITY OF SOLITUDE

In the end, when we continued to be stuck in the Háldijávri cabin, there was nothing left to say. Although we were surrounded by people, everyone retreated more and more into themselves as the hours passed. We were slowly but surely becoming part of the timeless ancient mountains that surrounded us. It was as if lichen would soon come to consume us, too, much the same way it had claimed the forsaken shoe on the fjell. Our minds drifted this way and that to places beyond the here-and-now. We were there together, for sure, and we acutely acknowledged that, yet somehow the sense of a continued isolation (however brief in absolute terms) pushed us all into our own inner worlds.

Although our experiences were not exactly spiritual, or at least not religious, each of us could readily understand how isolation can generate spiritual experiences. Such experiences are first and foremost about an awareness of how we are entangled with a reality extending beyond the world as it readily presents itself to us (e.g. Greenwood 2009). Isolation can help us to "tune up" to reality differently. So too can modern technology, as evidenced by techno-paganism. Contemporary information and communication technology can also present our world and our being in the world, including spatialities and temporalities, differently from what has conventionally been considered as the "real" world, and hence generate a sense of enchantment and spiritual experiences (e.g. Aupers 2009).

As for everyday human connections, our isolation and auto-ethnographic sensibilities also made us aware of our group dynamics and ways of relating with others, with the balances between sociality and solitude for everyone varying situationally. Roger and Oula are, at least outwardly, the most sociable ones in our group of five, and throughout our days together it was very difficult for anyone to be truly in solitude; the difficult weather effectively bound us together, as it was not really feasible (or at least particularly enjoyable) to wander off on one's own for any longer period of time. Markus and Aki had some solitude in their own tents on Ritničohkka. Still, especially down in Háldijávri, it was evident that everybody was anxious in their turn to escape the company of others, be it friends or new acquaintances.

In the "real world", Roger is constantly communicating online: sending messages on WhatsApp or other platforms, recording voicemail messages, writing emails, participating in meetings remotely and making calls on the telephone—often doing several of these things simultaneously. In the Ritničohkka cabin, he was constantly checking outside for an internet connection in the cabin, knowing that there was a fleeting and unruly chance of getting a momentary connection. To be fair, the rest of us also went out into the storm frequently to connect, however briefly, to the "ordinary" world outside Ritničohkka. But Roger clearly suffered from the worst "online hangover" among us—until, that is, we relocated to Háldijávri, where there was absolutely no chance of a phone or internet connection. With all ties to his electronically expanded and super-boosted everyday world cut—and nothing whatsoever he could do about it—Roger eventually became relaxed and restful in an entirely new way.

Being in nature, away from the sounds and sights of "modern" life, creates a physiological sensation that has been termed "soft fascination" by environmental psychologist Stephen Kaplan (1995). In such a state, Kaplan argues, we feel simultaneously transported, calm and buoyant—something one might experience while gazing at a handsome sunset or pausing in the middle of a forest trail to gaze at the sprawling view below. For Kaplan, spending time in nature helps restore our normal cognition states when these are overstimulated or over-taxed by directed attention. Through nature, it is said, the mind is allowed to drift, gaze, wander and be immersed in the moment, taking in experience through multiple senses—sights, sounds, smells, sensations—rather than working hard to filter things out so as to be more productive. Indeed, the primary reason people seek out solitude in wilderness spaces is to de-tether from the digital or disconnect from their devices (Long et al. 2003). Humans may in fact be less likely to feel lonely and more likely to feel "connected" in a spiritual sense when they are alone in nature—rather than being at home by themselves or alone in a public place.

In his work *Landscapes of the Mind*, John Porteous (1990) proposes that scientists and travellers alike should return to so-called groundtruthing as a fundamental method of exploration. He calls the method "intimate sensing", separating it from remote sensing by referring to the latter as clean, cold, detached and easy, whereas the world—the very world we want to explore, experience and understand—is seen as chaotic and messy. Intimate sensing, on the other hand, is involved, warm and rich, providing rewards beyond the domain of the intellect. This mode of intimate sensing de facto comprises a sense of disconnection, of disconnecting. The notion of digital disconnection is a critical response to the mediated conditions that characterize our societies and permeate our everyday lives (Lomborg and Ytre-Arne 2021).

At the same time as global society is becoming increasingly mediated—saturated by the digital media and infrastructures that now co-constitute our social worlds—so too has the notion of digital disconnection seeped into new domains of social life (Hepp 2020; Moe and Madsen 2021). We can understand such disconnection alongside notions of health, concentration, existentiality, freedom and sustainability.

As a group, we were five people affiliated with three different institutions and representing two different disciplines and three different nationalities. We all had our own distinct scientific reasons for wanting to be part of the journey to Ritničohkka, in addition to distinct social reasons for wanting to be away for a week with other people, as well as particular reasons for wanting to be in a part of the world that is more rural and remote than where we each resided at the time.

The plans with which we began our journey were rather open and flexible, so that each member of the group was able to carry out either some natural science or humanities-related task on their own, or science-journalistic observation, either socially as part of the group or in solitude. Our daily tasks shifted between these spheres. Social "group tasks" included, for instance, working together to dismantle the window frames of the cabin to enter it in the first place, constructing a makeshift oven outside for cooking dinner, and walking together—or at least in the vicinity of one another—on the fjell to make sure that nobody got lost or injured. Individual tasks included foraging the storage cabin for firewood, writing field diaries, daily personal ablutions and individual scientific documentation tasks, such as cataloguing found reindeer bones and recording the find locations.

Overall, this book was written much the same way as the fieldwork was done—together and in solitude. Some two-thirds of the material for the book was produced during two writing retreats in the tranquillity of a beachside pavilion in Helsinki, Finland, overlooking the frozen sea. While we had frequent planning and discussion sessions during those intensive writing periods, most of the time we were working independently on different parts of the book, generating text as well as editing and rewriting our own and each other's material, working in a kind of body-mirroring mode, in sort of subconscious, limbic resonance with each other (Scioli and Biller 2009: 154).

Somehow, we all mostly seemed to understand what needed to be done and when—or, at least, it was quite easy to arrive at a consensus. There were no fights among any of us, even if there might have been some latent frustrations which were not voiced and had to do with everybody's own little eccentricities. Markus and Oula had secured the funding for the research and were responsible for carrying it out as planned and delivering what was promised, whereas Aki, Roger and Vesa-Pekka were dedicated to doing any tasks assigned to them; at the same time, they experienced the place and work from a rather "free" perspective, and hence were perhaps attuned to seeing and thinking about things that might otherwise have gone unnoticed.

We talked about these group dynamics both during and after the fieldwork—again, as per our previous interests in what actually happens during fieldwork. A lot does happen and it can be difficult to verbalize. Knowing this, during our last writing retreat, Roger attempted a kind of an "automatic writing" experiment to capture this (Fig. 7.8). Working on this present final section, Roger, whose mind was swarming with ideas faster than he could write them, dropped his forehead on the desk and continued writing without looking at the screen, like a medium channelling a possessed spirit. He seemed to be on to something, but, perhaps befitting what the weird and monstrous is about, it is difficult to tell what exactly it all amounts to. Here is a small sample of what he wrote during his "trance":

> What did each of us get out of the journey, on a personal level? Was the solitary nature of being in this remote and quiet and off-the map space a goal in itself? Was this experience mitigated by us having a group? Indeed, the space itself, and the journey that was required to get to the mountain, proved enough of a defamilarization with our usual norm that once out there, there is nothing else to accept than what is in front of you.
> Even as we sat down to write the book together, four of us gathered in a large room in Helsinki, looking out to the western bay, and we are all sitting in silence either listening to music or to the tapping of keyboards, while writing at individual tables and our laptops, in solitude yet working on a shared document. Even then we were definitely working as a team, even if we largely wrote our own individual blocks of text.
> It was perhaps the fact that we were in a group with others that pushed us more and more to find the space just outside the deadelinzzone. Had one been on one's own, it might have been easier with such cans.

The walk from Moldy to Halti took the better part of a day, about 3\hours in total through . The weather that day was not the best, with fog some light rain and a few flurries of snow. But we made it all the way from one to the other cabin. The walking itself was, like the other parts of the trip, one of both individual solitude and group commenting: telling the joke of one of the s-caled everything

In the other cabin, once the Puukarjalaiset had and other hikers in distress had been carried off by the helicopter, we were forced to take stock of where we were, and with whom.

There was one cabin building in which there were separate sides accommodation and living spaces. Ours was filled with xxx (who?); the second had in it a motor and daughter duo from soher than we really wanted to. The In each of the cab bted for the next several days, and there was indeed the risk we were all to be stranded in the cabin together, with the knowledge hat we might well really have to get to know each other bether had been forecasgoing to be spending at least the next 12 to 18 hours (more) in the cabin together ... potentially more in fact because inclement weatelf-deprecating moment of sense of humor, which was unodubetedlhy ainmned to bring the speaker closer together to the listener – with the knowledge that we were likely stuck.

not outside of our pre-formed groups, although several people in the cabin di offer to share snacks or whiskey with others. The alcohol's indeed one aspect that enabled or facilitated social connectivity with unknown peoples. Why was this?

Ou.

The connective nature of social periods / periods of sociality around or fueled by alcohol is an interesting one for reflection. What was it about sharing, or at least offering to share, such a precious resource withsoneone you didn't know, had just met, would likely never see again, and certainly had no more or social obligation to share with?

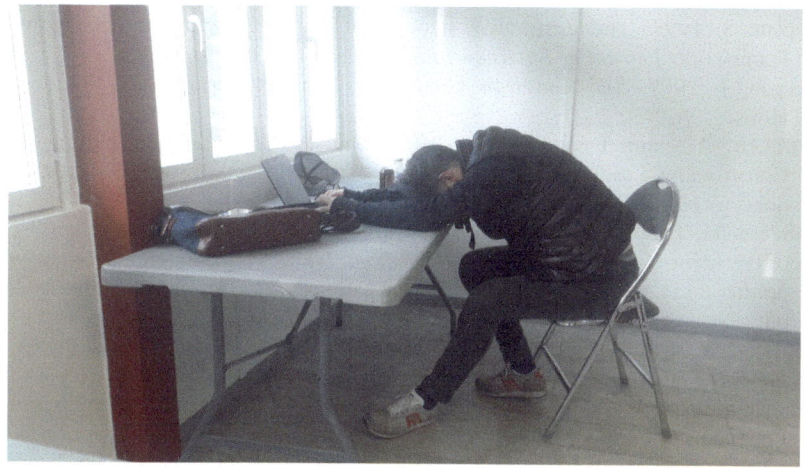

Fig. 7.8 Roger engaged in "automatic writing" during a writing retreat for this book in Helsinki

References

Aupers, S. 2009. "The Force Is Great": Enchantment and Magic in Silicon Valley. *Masaryk University Journal of Law and Technology* 3 (1): 153–173.

Burhaein, E., N. Demirci, D.T.P. Phytanza, A. Nadzalan, and E. Niksic. 2024. Is Walking a Miracle Cure for Active and Healthy Aging? *International Journal of Active & Healthy Aging* 2 (1): 10–17.

Doner, J. 2022. The Dynamic Uncertainty of Narrative, Place, and Practice in Spiritual Experience: Clues from the Phenomenology of Walking a Labyrinth. *Archive for the Psychology of Religion* 44 (3): 129–146. https://doi.org/10.1177/00846724221131149.

Dönmez, S. 2018. Hybrid Beings and Representation of Power in the Prehistoric Period. *Tarih Araştırmaları Dergisi* 37 (64): 97–124. https://doi.org/10.1501/Tarar_0000000695.

Erkkilä, V., and P. Iivari. 2019. *Kylmää sotaa Lapissa*. Helsinki: Otava.

Greenwood, S. 2009. *The Anthropology of Magic*. Oxford: Berg.

Gwynne, S.C. 2014. *Rebel Yell: The Violence, Passion, and Redemption of Stonewall Jackson*. New York: Simon and Schuster.

Hepp, A. 2020. *Deep Mediatization*. London: Routledge.

Ingold, T. 2004. Culture on the Ground: The World Perceived Through the Feet. *Journal of Material Culture* 9 (3): 315–340.

Kaplan, S. 1995. The Restorative Benefits of Nature: Toward an Integrative Framework. *Journal of Environmental Psychology* 15 (3): 169–182.

Keegan, J. 2009. *The American Civil War*. London: Vintage UK, Random House.

Kuronen, T., and J. Virtaharju. 2015. The Fishing President: Ritual in Constructing Leadership Mythology. *Leadership* 11 (2): 186–212. https://doi.org/10.1177/1742715013515697.

Lähdesmäki, T. 2009. From Personality Cult Figure to Camp Image—The Case of President Urho Kekkonen. *Particip@tions. Journal of Audience & Reception Studies* 6 (1): 52–76. http://www.participations.org/documents/lahdesmaki.pdf.

Länsman, A.-S. 2004. *Väärtisuhteet Lapin matkailussa Kulttuurianalyysi suomalaisten ja saamelaisten kohtaamisesta*. Inari: Kustannus-Puntsi.

Lohmann, R., and R.J. Letcher. 2023. The Universe of Fluorinated Polymers and Polymeric Substances and Potential Environmental Impacts and Concerns. *Current Opinion in Green and Sustainable Chemistry* 41: 100795.

Lohmann, R., I.T. Cousins, J.C. DeWitt, J. Glüge, G. Goldenman, D. Herzke, A.B. Lindstrom, M.F. Miller, C.A. Ng, S. Patton, M. Scheringer, X. Trier, and Z. Wang. 2020. Are Fluoropolymers Really of Low Concern for Human and Environmental Health and Separate from Other PFAS? *Environmental Science & Technology* 54 (20): 12820–12828.

Lomborg, S., and B. Ytre-Arne. 2021. Advancing Digital Disconnection Research: Introduction to the Special Issue. *Convergence* 27 (6): 1529–1535. https://doi.org/10.1177/13548565211057518.

Long, C.R., M. Seburn, J.R. Averill, and T.A. More. 2003. Solitude Experiences: Varieties, Settings, and Individual Differences. *Personality and Social Psychology Bulletin* 29 (5): 578–583.

McDougall, C. 2009. *Born to Run: A Hidden Tribe, Superathletes, and the Greatest Race the World Has Never Seen*. New York: Random House.

Moe, H., and O.J. Madsen. 2021. Understanding digital disconnection beyond media studies. *Convergence* 27 (6). https://doi.org/10.1177/13548565211048969.

Naum, M. 2016. Between Utopia and Dystopia: Colonial Ambivalence and Early Modern Perception of Sápmi. *Itinerario* 40 (3): 489–521.

Nenonen, M. 1999. Juokse sinä humma. In *Maata, jäätä, kulkijoita: tiet, liikenne ja yhteiskunta ennen vuotta 1860*, ed. T. Mauranen, 143–367. Helsinki: Tielaitos.

Porteous, J.D. 1990. *Landscape of the Mind: Worlds of Sense and Metaphor*. Toronto: Toronto University Press.

Ranta, J. 2023. Kaarina Karin vaelluskansiot – Suomen retkeilyhistorian pioneerin Haltin valloitus. *Retki*, 26 August. https://retkilehti.fi/retket/kaarina-karin-vaelluskansiot-suomen-retkeilyhistorian-pioneerin-haltin-valloitus/.

Scioli, A., and H. Biller. 2009. *Hope in the Age of Anxiety*. Oxford: Oxford University Press.
Seume, J.G. 1806. *Mein Sommer 1805*. Leipzig: Steinacker.
Suomi, J. 1988. *Vonkamies: Urho Kekkonen 1944–1950*. Helsinki: Otava.
Thoreau, H.D. 1862. Walking. *The Atlantic Monthly, A Magazine of Literature, Art, and Politics*. Boston: Ticknor and Fields IX (LVI) (June): 657–674.
Valli, S. 2018. Kääntäjän alkusanat. In J. G. Seume [1806], *Kesä 1805: matka Saksasta Baltian ja Venäjän halki Suomeen, Ruotsiin ja Tanskaan*. Helsinki: Osuuskunta Poesia.
Valtonen, T. 2019. Miten Saanasta tuli pyhä? Erilaisten rinnakkaisten Saanadiskurssien tarkastelua. *Terra* 131 (4): 209–222.
Veijonen, P. 2022. Olen päättänyt luopua "valloitus" sanan käytöstä. https://www.wildadventuresnorth.fi/l/olenpaattanut/, posted 10 November 2022.
Vergunst, J.L., and T. Ingold. 2008. *Ways of Walking: Ethnography and Practice on Foot*. Abingdon: Routledge.
WHO. 2022. *Global Status Report on Physical Activity 2022*. New York: World Health Organization.
Wiley, B.I. 2004 [1943]. *The Life of Johnny Reb. The Common Soldier of the Confederacy*. Baton Rouge: Louisiana State University Press.
Wuokko, M. 2011. Sport, Body and Power: Reassessing the Myth of President Kekkonen. *Norma: International Journal for Masculinity Studies* 6 (2): 124–140. http://www.idunn.no/ts/norma/2011/02/art07.
Yusha, Z.M. 2024. The Tuvan Depictions of Shamanic Headdresses. *Tomsk Journal of Linguistics and Anthropology* 3 (45): 113–122. https://doi.org/10.23951/2307-6119-2024-3-113-122.

Open Access This chapter is licensed under the terms of the Creative Commons Attribution 4.0 International License (http://creativecommons.org/licenses/by/4.0/), which permits use, sharing, adaptation, distribution and reproduction in any medium or format, as long as you give appropriate credit to the original author(s) and the source, provide a link to the Creative Commons license and indicate if changes were made.

The images or other third party material in this chapter are included in the chapter's Creative Commons license, unless indicated otherwise in a credit line to the material. If material is not included in the chapter's Creative Commons license and your intended use is not permitted by statutory regulation or exceeds the permitted use, you will need to obtain permission directly from the copyright holder.

CHAPTER 8

Conclusion: Monstrous Worlds

It turned out that there was little for us to do in the Háldijávri hut except for being, thinking, talking and observing our confined little world. This book largely derives from those conversations and from the experiences that preceded them. We have attempted to document, through an ethnographic approach encompassing both "fielding the field" and "fielding the mind", human-landscape relations in the context of archaeological fieldwork in a remote Arctic environment that is entangled with diverse historical and contemporary imageries and imaginaries. We have uncovered several aspects of meaning-formation emerging from the situational awareness of landscapes and engagements with them (Fig. 8.1).

People may not be consciously aware of how they "connect" to the environment and how the environment "connects" to people. Nonetheless, it is important to identify and recognize the affective qualities of landscapes because they mediate the ways people perceive and relate to them, irrespective of whether we are consciously aware of the affects. This applies to archaeological fieldwork, as well as everyday perceiving and relational knowledge of landscapes. Our experiences on Ritničohkka, as described and analysed in this book, thus ultimately contribute to an understanding of the dynamics of meaning-formation in human-landscape interaction, even though the specific meanings of landscape are, of course, highly contextual.

Ritničohkka very much underlined the significance of the uncertainty of the world and "being in the world". It substantiated a deeply dynamic

Fig. 8.1 Farewell to Ritničohkka, its two summits rising above the clouds, as seen from Háldi

and entangled nature, as well as the permeating yet elusive presence of invisible worlds and more-than-human agency. This is not exclusive to Ritničohkka and Lapland, however. Still, arguably, aspects of the lived world are simply more acutely sensed in the remote lands of the High North. This acuteness has evidently been experienced by generations of travellers. Moreover, in indigenous knowledge, these places and their more-than-human elements constitute the foundation for relational cosmologies.

The reason Ritničohkka presented itself to us as a weird otherworldly landscape was precisely because we came to accept it in relation to ourselves. Thus, a bodily-cognitive entanglement with the world around us embraced us in a way that differed from the impact of our familiar everyday environment. Through its embrace, we became differently conscious and aware of the surrounding world, its manifold dimensions, layering and dynamics.

Our presence *in* the landscape altered our faculties of perception and prompted theoretical reflections about the world and its Anthropocenic condition. There was more than the "monstrous"—that is, dismissive of humanity and its grand designs—to the fjell landscape. It also took on the roles of a wisened observer as well as a victim. The entire globe in the age of the Anthropocene is not unlike an enormous monster, revealing itself to us obscurely, as many scholars working on "weirding" (whether under this very banner or not) and the Anthropocene have suggested.

The contemporary world has myriad monstrous qualities, including various invisible forces that are by-products of human activity yet seemingly beyond control. Digital and other high modern technological networks have altered space–time relations in fundamental ways, akin to incorporeal masterminds operating in weird non-linear space–time domains. The lived world, whether metropolitan cities or the high northern "wilderness", is indeed full of invisible and hidden beings and powers, some of which we—partly unexpectedly—encountered and engaged with on Ritničohkka.

Increased connectivity is transforming the world into a global metropolis. With it, solitude itself is undergoing a change. Being social in an overpopulated world is a task which not everyone is geared up for. Solitude persists even in the most buzzing hubs, where alienation from society pushes people into their own homes, their own little worlds. Against this backdrop, the solitude in a desolate and remote place is exceedingly social; here, even a solitary person, while separated from other human actors, is deeply enmeshed with the surrounding world. This occurs through a close connection not only with geoforms and plant-life, but also with the material networks and the hive mind of the world-wide marketplace and its literal encumbrances. A global metropolis, encroaching on proverbial untouched wilderness: a monstrosity in the making.

We all live in our own time, a time whose future is unknown. All we can do is interpret trajectories. For a long time, the narrative of population growth was undeniable. In the so-called developed countries, this myth has already broken down, with population decline either already taking place or predicted to trend in the near future. The overall world population has continued to grow for decades. But a startling projection of population growth published in 2023 suggests that not only will the world population peak during the twenty-first century, possibly in the 2080s, but that this will be followed by a population crash as dramatic as the growth that began in the early industrial period several centuries ago.

According to this model, after passing the 10 billion mark, the global human population will decline significantly in the three centuries that follow, settling at a level not seen since the post-glacial Stone Age (Spears et al. 2023). The crash predicted in these projections is not caused by resource depletion, famine, disease or the climate crisis, but solely by declining fertility due to rising living standards.

As bad as humans are at predicting the future, this projection reframes the Anthropocene. There is no doubt that humanity has become a geological actor, and designating our own time as a new epoch is not merely human-centric or present-centric (e.g. Halliday 2022: 285–303). Humans have drilled into bedrock and mined through millions of years of stratigraphy. We have hollowed out pit coal deposits, the mass graves of the rainforests of the Carboniferous Period, and released their atoms into the stratosphere after 300 million years of dormancy. Nuclear tests from the 1950s have caused atmospheric carbon isotope configurations to desynchronize from the past. It is estimated that, in 2020, the total mass of anthropogenic material (mostly concrete and gravel) exceeded the total biomass of the world, with continued exponential growth heading towards a 200% increase in the next sixty years (Elhacham et al. 2020). This global population and anthropogenic production peak will be represented in the geologic record, akin to the Cambrian explosion that occurred over 500 million years ago, which is represented in the bedrock of Kilpisjärvi.

What will twenty-third-century archaeology be like? Will our own population explosion overshadow everything that precedes it, with prior material cultures fading out, as postulated by the authors of the Silurian hypothesis (Schmidt & Frank 2018)? This hypothesis explores the future record of the Anthropocene and is named after a reptilian civilization that predates humanity in the sci-fi series *Doctor Who*. The fictitious species was named for its origin in the geological period 420 million years ago— which, incidentally, also gave life to the stones of Ritničohkka. It is strange how such sci-fi stories, as well as Lovecraftian weird fiction, can easily hark back millions of years, yet looking forward—even just a few centuries—is like wading through a storm inside an unyielding veil of cloud.

Perhaps Lapland and the Arctic sphere will, after a short burst of intense human activity—a sprawl of the global metropolis—return to the solitude and remoteness we expected of it. A (re)turn from one otherworld to another.

8 CONCLUSION: MONSTROUS WORLDS   191

Pondering the monstrous implications of our time and its relation to an unimaginably deep time gave rise to a new sense of otherworldliness in us. The question remains: what is otherworldliness in practice? One of our goals here was to reach some kind of understanding of the concept. But can otherworldliness really be defined?

We have examined many acute instances in which engaging with permeable borders makes categories lose their meaning. This is true not only in relation to classifications but also to time itself: the flattening of time, hundreds of millions of years compressed physically into stone, leading to disorientation brought forth by the perception of an intangible place.

Imagine the spatio-temporality of a helicopter gliding in a narrow space between elevated land and a low cloud ceiling, over thousands of years of human activity residing in a post-glacial landscape of 400-million-year-old Silurian geology, recognizable from early twentieth-century black-and-white photographs and twenty-first-century Light Detection and Ranging (LiDAR) images.

Consider guiding a leather hiking boot with one's foot—a living organism inside the skin of a dead organism—next to a lone dead leather shoe that is slowly becoming part of the living lichen around it arranged into the shapes and features of a microcosmic landscape recognizable in scaled-up maps on a GPS navigator screen consulted in an open landscape lost inside the closed world of a cloud.

Staring at the light emanating from an iPad preloaded with Hollywood movies in a dark, mouldy, shadow-ridden cabin without heat or electricity—a technological wonder inside a technological failure where even a simple key refuses to work—transfixed by the screen, which transmits your sense of being into a fictional story full of colour and senses, then snapped back to the dull grey moment by a movement at the corner of your eye, in the shadow, where nothing could possibly have moved.

Searching for reindeer bones in an endless rocky landscape and happening to pick up a thumb-sized prehistoric stone artefact, a meaningful rock from two millennia ago, among a mountain of millions of seemingly inconsequential rocks that carry hundreds of millions of years of narrative.

Attuning an enminded body to weirding, and allowing the landscape to seep through it, transducing the flow into unexpected realms, while being critically mindful of exoticism and its colonialist tendencies, even as you accept that you are also a product of them.

Breaking down barriers of the self and the surrounding world, the living and the inorganic, imagining the vastness of time in relation to a moment. All sense of reference—of fixed points of relations, classifications and categories, ontologies, definitions and delimiters—melts down and trickles into the dark cavities between the rocks of the endless stone sea.

Otherworldliness and its monstrosities, then, are defined by their indefinability.

References

Elhacham, E., L. Ben-Uri, J. Grozovski, Y.M. Bar-On, and R. Milo. 2020. Global Human-Made Mass Exceeds All Living Biomass. *Nature* 588: 442–444.

Halliday, T. 2022. *Otherlands: A World in the Making*. Dublin: Penguin Random House.

Schmidt, G.A., and A. Frank. 2018. The Silurian Hypothesis: Would It Be Possible to Detect an Industrial Civilization in the Geological Record? *International Journal of Astrobiology* 18 (2): 142–150. https://doi.org/10.1017/S1473550418000095.

Spears, D., S. Vyas, G. Weston, and M. Geruso. 2023. Long-Term Population Projections: Scenarios of Low or Rebounding Fertility. September 1. Available at SSRN: https://ssrn.com/abstract=4534047 or https://doi.org/10.2139/ssrn.4534047.

Open Access This chapter is licensed under the terms of the Creative Commons Attribution 4.0 International License (http://creativecommons.org/licenses/by/4.0/), which permits use, sharing, adaptation, distribution and reproduction in any medium or format, as long as you give appropriate credit to the original author(s) and the source, provide a link to the Creative Commons license and indicate if changes were made.

The images or other third party material in this chapter are included in the chapter's Creative Commons license, unless indicated otherwise in a credit line to the material. If material is not included in the chapter's Creative Commons license and your intended use is not permitted by statutory regulation or exceeds the permitted use, you will need to obtain permission directly from the copyright holder.

Correction to: Weirding Landscapes

Correction to:
V. Herva et al., *Weirding Landscapes,* **Arctic Encounters,**
https://doi.org/10.1007/978-3-031-85016-5

The original version of the book was inadvertently published without including the series forward which has now been included now and with an incorrect series editor affiliation as Roger Norum, Environmental Humanities, University of Oulu, Oulu, Finland which has now been updated to Roger Norum, Cultural Anthropology, University of Oulu, Oulu, Finland. The book has been updated with the changes.

The updated version of this book can be found at
https://doi.org/10.1007/978-3-031-85016-5

© The Author(s) 2025
V. Herva et al., *Weirding Landscapes,* Arctic Encounters,
https://doi.org/10.1007/978-3-031-85016-5_9

Open Access This chapter is licensed under the terms of the Creative Commons Attribution 4.0 International License (http://creativecommons.org/licenses/by/4.0/), which permits use, sharing, adaptation, distribution and reproduction in any medium or format, as long as you give appropriate credit to the original author(s) and the source, provide a link to the Creative Commons license and indicate if changes were made.

The images or other third party material in this chapter are included in the chapter's Creative Commons license, unless indicated otherwise in a credit line to the material. If material is not included in the chapter's Creative Commons license and your intended use is not permitted by statutory regulation or exceeds the permitted use, you will need to obtain permission directly from the copyright holder.

Correction to: The "Glacier"

Correction to:
Chapter 3 in: V. Herva et al., *Weirding Landscapes*, Arctic Encounters, https://doi.org/10.1007/978-3-031-85016-5_3

In the original version of this book, several passages at the end of Sect. 3.4 have been amended to include proper attribution to source material that was inadvertently unacknowledged. This occurred as a result of an oversight during the writing and editing process. The book has been updated with the changes.

The updated version of this chapter can be found at
https://doi.org/10.1007/978-3-031-85016-5_3

© The Author(s) 2025
V. Herva et al., *Weirding Landscapes*, Arctic Encounters,
https://doi.org/10.1007/978-3-031-85016-5_10

Open Access This chapter is licensed under the terms of the Creative Commons Attribution 4.0 International License (http://creativecommons.org/licenses/by/4.0/), which permits use, sharing, adaptation, distribution and reproduction in any medium or format, as long as you give appropriate credit to the original author(s) and the source, provide a link to the Creative Commons license and indicate if changes were made.

The images or other third party material in this chapter are included in the chapter's Creative Commons license, unless indicated otherwise in a credit line to the material. If material is not included in the chapter's Creative Commons license and your intended use is not permitted by statutory regulation or exceeds the permitted use, you will need to obtain permission directly from the copyright holder.

Index

A

Absence, 10, 23, 26, 63, 91, 95
Academic, 3, 22, 33, 115, 116
Access, 7, 8, 52, 53, 105, 142, 144, 154, 155
Action, 35, 61, 62, 71, 113, 144
Air, 47, 49, 93, 103, 111, 142, 143, 156
Ancient, 4, 25, 64, 65, 68–71, 87, 91, 107, 109, 163, 175, 178
Animal, 10, 53–55, 57, 60, 71, 90–92, 95, 108, 117, 135, 137, 142, 169–171
Anthropocene, 4, 6, 10, 11, 16–22, 24–26, 31, 32, 68, 72, 82, 102, 130, 141, 189, 190
Anthropology, 29, 45
Archaeological, 6, 9–11, 22–24, 28–30, 33, 35, 45, 56, 91, 117, 120, 131, 132, 134, 137, 167, 172, 176, 187
Archaeological research, 35
Archaeological survey, 6, 7, 11, 29, 47, 134

Archaeology, 23, 24, 30, 33, 35, 110, 117, 136, 137, 190
Arctic, 4, 6, 7, 16, 20, 45, 88–90, 132, 134, 142, 172, 187, 190
Arctic landscapes, 19, 35
Arctic tourism, 4
Area, 6, 33–35, 46, 47, 49, 51, 52, 54–56, 63, 84, 90, 91, 105, 113, 117, 131, 133, 136, 155, 156, 159, 171, 173, 176
Artefact, 70, 108, 135, 137, 163, 191
Aviation, 49

B

Basecamp, 104, 134, 174
Being, 7, 17, 18, 20, 23, 25–28, 32, 35, 36, 49, 52, 59, 60, 68, 72–74, 82, 85–91, 94, 95, 100, 102, 106, 107, 113–121, 130, 135, 138, 140, 141, 143, 145, 155, 157, 162–164, 169, 178, 179, 181, 187, 189, 191
Biodiversity, 60
Bodies on the Move, 27

Body, 23, 26, 60, 92, 137–139, 141–143, 156, 161–163, 167, 168, 180, 191
Border, 47, 49, 52, 54, 86, 105, 109, 145, 191
Border guard, 92, 99, 101, 155, 156, 175, 176
Boulder, 27, 48–50, 55, 56, 59, 64, 65, 69, 72, 73, 82, 83, 85, 90, 91, 110, 115, 116, 128, 130, 169, 176
Breathing, 53, 89, 110, 136

C
Cabin, 2, 28, 49, 50, 55, 92, 94, 99–104, 106, 107, 109–115, 118–121, 135, 136, 143, 153, 155–157, 159, 164, 170, 172, 174, 176, 178–180, 182, 191
Change, 7, 17, 22, 33–35, 55, 57, 61, 62, 86, 88, 90, 103, 112, 144, 146, 160, 165, 173, 189
Christmas, 4
Chthulucene (Haraway, Donna), 19
Circumpolar, 139
City, 6, 83, 106, 128
Climate, 6, 22, 31, 60, 61, 148, 190
Climate change, 7, 11, 17, 18, 22, 25, 31, 32, 61, 62
Cloud, 2, 3, 51, 87, 94, 106, 110, 111, 113, 116, 118, 127–130, 143–145, 148, 157, 162, 164, 190, 191
Communication, 106, 108, 109, 114, 121, 138, 178
Company, 48, 106, 107, 157, 158, 178
Conflict, 33, 158, 160
Conquest, 173, 174
Contemporary, 10, 18, 22, 25, 29, 72, 86, 105, 106, 109, 133, 178, 187

Contemporary world, 32, 189
Context, 7, 10, 11, 19, 20, 26, 29, 31–33, 35, 55, 72, 84, 94, 117, 131, 132, 134, 137, 139, 141, 154, 163, 170, 187
Cosmology, 108, 139
Crisis, 31, 165, 190
Cthulhu, 24, 68, 82, 83, 85
Culture, 10, 26, 35, 55, 71, 84, 85, 87, 108, 115, 121, 135, 138, 142, 154, 169, 173, 190

D
Death, 18, 61, 62, 71, 90, 91, 93–95
Decade, 18, 34, 35, 45, 61, 63, 102, 105, 107, 119, 141, 189
Deep past, 65, 70, 87
Development, 17, 35, 88, 105, 161
Dimension, 10, 17, 23, 26, 28–30, 66, 70, 85, 87, 88, 111, 114, 120, 121, 128, 130, 133, 135, 137, 139–141, 160, 170, 188
Discourse, 7, 166, 173, 174
Discovery, 4, 7, 66, 91, 130, 170
Disorientation, 48, 133, 134, 147, 164, 191
Diversity, 65, 89, 90, 135
Domestication, 142
Dynamic, 10, 11, 22, 23, 34–36, 115, 130, 140, 147, 178, 181, 187, 188

E
Emotion, 20, 23, 129, 141
Encounter, 16, 23, 27, 28, 30, 32, 34, 91, 120, 121, 130, 131, 145, 147, 148, 157, 158, 173
Engagement, 20, 28–30, 45, 87, 90, 121, 133, 136, 138, 164, 165
Environment, 6, 11, 20, 22, 23, 29, 32, 34, 36, 54, 60, 61, 71, 81,

85, 86, 89, 90, 92, 100, 108, 109, 113, 118, 121, 127, 129, 130, 136, 138, 140, 143, 144, 146, 154, 156, 159, 160, 187, 188
Escape, 87, 93, 95, 106, 164, 166, 178
Europe, 4, 29, 52, 65, 83
European, 7, 72, 94
Evolution, 89
Experience, 7, 19–23, 30, 32, 35, 36, 48, 55, 57, 59, 65, 73, 85, 86, 105, 106, 119, 130, 132, 133, 136–138, 141–143, 145, 148, 153, 157, 160, 165, 169, 172, 173, 178, 179, 181, 187
Extraction, 15, 27
Extractive, 17, 69
Extractive industry, 17
Extraordinary, 16, 22, 92, 157
Extreme, 34, 36, 160

F
Fennoscandia, 4, 65, 69
Fiction, 4, 17–19, 22, 24, 68, 86, 114, 119, 140
Field, 2, 3, 8, 20, 30, 33–35, 47–49, 53–56, 59, 64, 69, 73, 82, 83, 85, 90, 110, 115, 116, 128, 130, 133, 137, 143, 161, 164, 167, 169, 170, 180
Fielding, 10, 11, 32, 34, 35, 120, 130, 131
Fieldwork, 4, 6, 10, 11, 15, 20, 22, 23, 28–32, 34, 35, 45, 46, 48, 57, 85, 92, 120, 130–132, 143, 146, 172, 180, 181, 187
Film, 86, 112, 113, 118, 119, 127, 133, 167
Finland, 3, 5, 6, 9, 30, 33, 46–49, 51–53, 57, 64, 65, 70, 105, 115, 117, 132, 135, 144, 154, 158, 165–167, 173, 174, 176, 180
Finnish, 6, 8, 32, 33, 47, 52, 54, 56, 71, 84, 92, 95, 107, 109, 110, 115, 132, 134, 139, 148, 154–156, 158, 166, 167, 172–174, 176
Finnish Lapland, 1, 4, 6, 30, 64
Fjell, 3, 6, 8, 22, 47, 49, 51, 52, 59, 63, 70, 82, 84, 85, 87, 90–93, 95, 104, 107, 109–113, 115, 118, 127, 132, 133, 142, 146, 155, 167, 173, 174, 178, 180, 189
Flying, 49, 145
Foot, 48, 94, 147, 165, 166, 168, 170, 191
Forest, 5, 55, 118, 134, 155, 167–169, 179

G
Gear, 1, 2, 47, 145, 159–162, 164, 167
Geography, 30, 71
German, 53, 94, 117, 165, 166, 178
Glacial, 60, 69–71, 85, 86
Glacier, 9, 22, 27, 31, 45–47, 54, 56, 57, 59–63, 65, 69, 71, 72, 85, 134
Glacier extinction, 31

H
Háldi, 49, 52–54, 65, 69, 132, 147, 153, 159, 172–174
Hallucinatory, 157
Hannukainen mining site, 15
Haraway, Donna, 19
Helicopter, 1–3, 47–50, 54–56, 83, 85, 104, 118, 135, 143–145, 148, 159, 162, 164, 173, 182, 191

Herding, 29, 54, 55, 92, 93, 108, 171, 173
Heritage, 24, 29, 33, 53, 165
High North, 4, 45, 47, 73, 82, 90, 109, 119, 148, 154, 174, 188
High northern, 5, 28–30, 85, 90, 153, 189
Hiking, 3, 49, 109, 115, 154, 158–167, 169, 170, 173, 176, 191
Historical, 11, 29, 45, 73, 84, 94, 109, 166, 187
History, 19, 33, 56, 60, 83, 121, 131, 155, 156, 170, 173, 174
Horror, 4, 15, 17–19, 22, 25, 26, 68–70, 72, 73, 85, 86, 90, 110, 118–120, 127, 133, 140, 160
Huiputus, 172
Human, 6, 10, 11, 15–19, 22–26, 29–31, 34–36, 45, 48, 52–55, 60–62, 68, 69, 71, 73, 81–90, 92–95, 108, 113, 117, 119–121, 130, 136–139, 141, 142, 145, 153, 155, 157, 167–170, 173, 178, 179, 187–191
Humanities, 19, 180
Hut, 3, 54, 144, 147, 148, 153–160, 164, 175, 187

I
Ice, 7–9, 27, 54, 56, 57, 59–63, 65, 69–71, 82, 83, 85, 87, 118, 129, 133, 163
Ice and Snow, 52, 54, 59
Idea, 4, 15, 20, 21, 28–32, 34, 35, 45, 46, 57, 70, 71, 83, 85, 87, 91, 108, 110, 114, 118, 132, 135, 136, 138–140, 154, 166, 181
Impact, 3, 8, 11, 31, 93, 95, 131, 136, 188
Incorporeal, 134, 143, 144, 189

Industry, 170
Instrument, 23, 139, 141, 147
Invisible, 2, 11, 17, 22, 23, 26, 28, 53, 83, 86, 90, 91, 95, 102, 113, 119, 139–141, 143, 146, 188, 189
Isotope, 190

K
Key, 1, 2, 32, 90, 99, 116, 118, 135, 158, 191
Kilpisjärvi, 1, 33, 46–49, 52, 54, 70, 87, 104, 109, 132, 147, 148, 159, 173, 174, 190
Knowledge, 7, 8, 21, 23, 30, 33, 61, 70, 110, 131, 139, 140, 158, 166, 172, 182, 187, 188

L
Labyrinth, 132–134
Landscape, 1, 5–7, 10, 11, 15, 19, 20, 23, 27–30, 32–34, 36, 45, 46, 48, 49, 51–53, 55, 57, 59, 63–65, 68–71, 73, 74, 81, 83–87, 90–92, 94, 95, 107, 109, 110, 113, 115, 116, 119, 120, 129–135, 141, 144–146, 153, 160, 162, 164, 165, 171, 172, 174, 176, 187–189, 191
Lapland, 4–6, 16, 29, 32, 34, 45, 46, 53, 57, 65, 71, 82, 84, 85, 94, 95, 109, 117, 142, 145, 155, 157, 161, 165, 169, 172–174, 176, 188, 190
Legacy, vii
Life, 3, 10, 18, 22, 25, 26, 33, 34, 49, 65, 70, 73, 81, 87–91, 94, 95, 105, 106, 113, 119, 157, 173, 179, 180, 189, 190
Ligotti, Thomas, 18, 113

Living, 6, 17, 34, 47, 51, 60, 66, 84, 86, 87, 91, 99, 112, 119, 135, 156, 169, 173, 182, 190–192
Lovecraft, H.P., 4, 17–19, 22, 24–26, 68–71, 73, 82, 86, 87, 119, 128, 140, 141
Lovecraftian, 26, 68, 82, 86, 128, 141, 190

M
Maintenance, 92, 100, 101, 155
Materiality, 82, 136
Melting, 7, 9, 20, 22, 31, 46, 47, 52, 61, 63, 83, 92, 129
Memory, 156, 165
Mental, 1, 28, 133, 140, 143, 161, 170, 174
Mesolithic, 69, 135
Military, 107, 136, 174
Mindscape, 84, 95, 129, 132
Mineral, 64, 71, 87–89
Mobile, 1, 48, 49, 105–107, 113, 118, 136, 162, 170
Mobile technology, 105
Mobility, 45, 47, 53, 54
Monsters, 11, 15–19, 22–27, 32, 68, 71, 73, 82, 86, 103, 118, 119, 132, 133, 141, 189
Monstrous, 4, 10, 16–19, 22–27, 71, 82, 86, 93–95, 113, 114, 119, 120, 156, 181, 189, 191
Mosquito, 54, 93–95
Mountain, 2, 4, 6, 9, 50–52, 56, 60, 64, 65, 70, 73, 83, 85–89, 92, 108, 120, 142, 143, 159, 169, 178, 181, 191
Movement, 7, 10, 27, 55, 90, 116, 131, 133, 134, 156, 157, 166, 191
Mythical, 6, 132, 172, 174, 176
Mythical resonances, 51

N
National, 115, 158, 167
Neolithic, 31, 107
Network, 90, 100, 107, 114, 136, 154, 155, 161–164, 169, 189
North, 3, 4, 7, 8, 33, 35, 47, 49, 51–54, 56, 64, 72, 94, 95, 136, 139, 147, 155, 160, 161, 165, 167, 169, 172
Northern land, 5, 85, 145, 154, 160, 174
Northernmost, 4, 52, 109, 155
Northern wilderness, 109

O
Otherworldly, 1, 23, 24, 26, 65, 68, 70, 85, 90, 108, 110, 113, 119, 133, 153, 160, 162, 188

P
Patch, 54, 56, 57, 59, 60, 90, 92, 109, 118
Perennial, 54, 153
Perennial ice, 69, 92, 95
Performed, 54, 106, 117
Pilot, 1, 3, 48–50, 104
Place, 6, 7, 10, 15–17, 20, 22, 23, 28, 32, 35, 36, 47, 53, 55, 57, 59, 60, 64–66, 68, 70, 71, 81, 82, 92, 95, 101–103, 105, 107, 108, 110, 113, 115–120, 128, 129, 131, 133–136, 140–143, 153, 156–158, 164, 166, 170, 173, 178–181, 188, 189, 191
Polar, 6, 31, 35, 61, 160
Political, 6, 21, 117, 174
Popular culture, 18, 24, 25, 86, 110
Prehistoric, 2, 29, 70, 109, 191
Presence, 10, 17, 23, 26, 32, 33, 59, 63–65, 69, 71, 74, 82, 85, 87,

90, 91, 94, 95, 107, 113, 115, 119, 137, 157, 188, 189

R
Reality, 1, 10, 19, 21–27, 32, 36, 49, 69–71, 113, 114, 117, 119–121, 129, 133, 134, 136, 138–142, 145, 146, 153, 159, 160, 178
Region, 6, 7, 31, 33, 34, 46, 51, 54, 60, 64, 65, 69, 70, 84, 87, 91, 108, 142, 154, 159, 164, 174
Reindeer, 5, 29, 33, 34, 45, 47, 48, 54–57, 90–95, 108, 110, 130, 132, 155, 169, 171, 173, 180, 191
Relation, 6, 10, 11, 16, 19, 20, 22, 26, 27, 29, 34, 35, 53, 87, 90–92, 99, 108, 113, 118, 129, 134–136, 140–142, 154, 157, 159, 165, 187–189, 191, 192
Relational ontologies, 10, 22, 136
Remote, 4, 6, 7, 10, 35, 36, 46, 65, 84, 85, 99, 101, 106, 112, 113, 119, 130, 143, 145, 154, 156, 158, 160, 164, 174, 179–181, 187–189
Remoteness, 59, 109, 190
Resonance, 28, 49, 91, 107, 110, 128, 180
Resource, 8, 166, 174, 182, 190
Ritničohkka, 4, 6, 8, 9, 11, 15–17, 22, 23, 26–32, 34–36, 45–49, 52–56, 59, 63–66, 69, 70, 73, 82–87, 90, 92, 95, 99, 103, 107–110, 113–115, 118–120, 127–135, 141–145, 147, 148, 153, 155, 156, 162, 164, 169, 170, 174, 176, 178–180, 187–190
Road, 51, 167, 176
Rock, 2, 27, 49, 50, 55–57, 63–66, 68–71, 73, 74, 81, 82, 87, 90, 91, 107, 108, 111, 113, 116, 118, 120, 127–129, 132, 134, 170, 191, 192
Rural, 105, 106, 180

S
Sámi, 4–6, 8, 27, 32–34, 45, 47, 53–55, 86, 91–94, 107–109, 118, 132, 134, 139, 169, 172, 173, 175
Sámi culture, 4
Sápmi, 4, 6, 34, 35, 172
Scandinavia, 64, 169
Second World War, 29, 53, 54, 94, 117, 166, 171
Sense, 4, 15–17, 19–29, 34–36, 45, 48–50, 55, 60, 65, 68–71, 73, 74, 81, 83, 85, 86, 90, 92, 95, 106, 109–111, 114, 119, 129, 132, 133, 136–139, 141–145, 153, 154, 156, 157, 160, 165–167, 170, 172, 178, 179, 182, 191, 192
Shadow, 1, 6, 135, 136, 156, 157, 191
Shamanism, 28, 139
Siida, 54, 55, 109
Sky, 1, 50, 64, 66, 72, 111, 118
Smell, 131, 142, 143, 179
Snow, 4, 7, 8, 51, 54, 56, 57, 59, 60, 63, 69, 90, 92, 118, 182
Sociality, 178, 182
Society, 35, 106, 115, 117, 164, 166, 173, 180, 189
Solitude, 178–182, 189, 190
Sound, 94, 110, 131, 136, 137, 145, 146, 148, 179
South, 15, 64, 109, 115, 132, 158, 165, 174, 176
Southern, 5, 47, 148, 155, 165, 173
Spectral, 108, 156, 157

Spiritual, 10, 23, 45, 63, 89, 107–110, 114, 128, 130, 133, 157, 178, 179
Stone, 2, 48, 59, 64–66, 71, 83–85, 89, 92, 107, 132, 134, 146, 176, 190–192
Story, 3, 36, 57, 83, 90, 93, 113, 128, 141, 191
Subterranean, 15, 70, 133
Summit, 52, 61, 95, 104, 107, 109, 112, 116, 144, 147
Surface, 1, 11, 15, 16, 47, 55, 59, 63, 65, 66, 70, 73, 82, 113, 116, 129, 130, 139, 140, 146, 156, 169, 171
Survival, 161
Sweden, 5, 46–49, 52, 53, 70, 132, 166, 175, 176

T
Tale, 64, 68, 69
Technology, 11, 36, 106, 108, 110, 114, 138, 139, 159, 163, 164, 168, 170, 178
Temporality, 146
Theory, 30, 68, 139, 146
Thing, 176
Time, 1–7, 9, 10, 16, 19, 21, 23, 25, 32, 34, 36, 45–47, 50, 51, 54–56, 59, 60, 64–66, 68–71, 73, 74, 81, 82, 86, 89–92, 95, 101, 102, 105, 108, 109, 112, 115, 117, 118, 127, 133, 137, 139, 140, 143–145, 157–159, 166, 172, 174, 178–181, 189–192
Tipping points, 18
Tornensis, Juha, 33, 54
Tourism, 4–6, 8, 29, 57, 94, 105, 106, 154, 155, 173, 174
Trail, 3, 47, 49, 109, 115, 154, 155, 158, 159, 166, 175, 179

Transition, 6, 48, 153, 168
Transitional landscapes, 63

U
Uncanny, 18, 20–22, 24, 63, 82, 114, 157, 172
Unhomely (unheimlich), 20

V
VanderMeer, Jeff, 18, 20
Village, 33, 47, 53, 109, 132, 148, 173, 176
Vulnerable landscapes, 7

W
Walking, 59, 90, 111, 128, 130, 131, 133, 134, 138, 153, 161, 164–169, 173, 180, 182
War, 166
Weird, 4, 11, 17–27, 32, 66, 68–70, 73, 82, 85, 87, 102, 108, 110, 112–114, 118–120, 128, 133, 136, 138, 141, 147, 153, 154, 167, 172, 181, 188, 189
Weird fiction, 4, 17, 18, 20, 25, 27, 32, 35, 69, 71, 140, 141, 190
Weirding, 11, 18–24, 27, 28, 30, 36, 102, 153, 157, 189, 191
Weird landscape, 29, 147
Wilderness, 2, 3, 15, 16, 51, 86, 105, 117, 118, 135, 144, 153–158, 160, 167, 170, 172, 173, 175, 176, 179, 189
Wilderness, Arctic, 5
World, 1, 4, 6, 9–11, 15–26, 28–30, 35, 47, 49, 53, 65, 66, 68–73, 82, 85, 87, 88, 90, 105, 106, 111, 113–115, 117, 119–121, 127, 129, 130, 133, 134, 136–142, 145, 146, 148, 153–155, 157–160, 165, 169,

170, 173, 174, 176, 178–180, 187–192

Z

Zone, 95, 104–107, 115, 153, 160, 161, 170

The manufacturer's authorised representative in the EU is Springer Nature Customer Service Centre GmbH, Europaplatz 3, 69115 Heidelberg, Germany. If you have any concerns regarding our products, please contact ProductSafety@springernature.com

Printed and bound by CPI Group (UK) Ltd, Croydon, CR0 4YY
26/03/2026
02078942-0002